Observing the
Caldwell Objects

GW00645419

Springer
London
Berlin
Heidelberg
New York
Barcelona
Hong Kong
Milan
Paris
Singapore
Tokyo

By the same author .

The Art and Science of CCD Astronomy
ISBN 3-540-76103-9

Software and Data for Practical Astronomers:
The Best of the Internet
ISBN 1-85233-055-4

David Ratledge

Observing the Caldwell Objects

With 223 Illustrations

Springer

Cover illustrations: Images by Christian Buil, Eric Thouvenot, Tim Puckett, Pedro Re, Steve Lee, Alex Richter and David Ratledge

ISBN 1-85233-628-5 Springer-Verlag London Berlin Heidelberg

British Library Cataloguing in Publication Data
Ratledge, David, 1945–
    Observing the Caldwell objects
    1.Astronomy – Observers' manuals   2.Astronomy – Charts,
    diagrams, etc.
    I.Title
    522
ISBN 1852336285

Library of Congress Cataloging-in-Publication Data
Ratledge, David, 1945–
    Observing the Caldwell objects / David Ratledge.
        p. cm.
    ISBN 1-85233-628-5 (alk. paper)
    1. Astronomy–Observers' manuals. 2. Astronomy–Charts,
diagrams. etc. 3. Galaxies–Charts, diagrams, etc. 4. Stars–Clusters–
Charts, diagrams, etc. 5. Nebulae–Charts, diagrams, etc. I. Title.
QB64.R37 2000
523–dc21   99-057027

Typeset by EXPO Holdings, Malaysia
Printed and bound at the University Press, Cambridge
58/3830–543210 Printed on acid-free paper SPIN 10732251

Contents

Introduction

Background

A new collection of deep sky targets for the modern amateur astronomer was long overdue. The traditional catalogue of deep sky objects, i.e. that produced by Messier over 200 years ago, has become a victim of its own success. Those objects have now become far too well known, over-viewed, over-imaged and over-published. More importantly they have often resulted in more interesting objects being totally overlooked. The logic behind the compilation of Messier's objects was nothing more profound than that they might be confused for a comet! There must be a better way to choose the best deep sky objects for the modern era. It was time for a fresh approach.

Enter Patrick Moore. As an active observer for over half a century and the last 40 years of which, as presenter of BBC TV's *Sky at Night*, he has travelled all over the World and never missed the opportunity to observe the skies. This has given him an in-depth, practical knowledge of deep sky objects in both northern and southern hemispheres. Drawing on this almost unrivalled experience, he has produced his own 109 object catalogue (ignoring the Messier objects) stretching virtually from pole-to-pole. Remember that the Messier 109 (or 110) objects are really for northern observers, being limited to declination 34°S. It therefore omits many Southern gems. In addition, the unique format devised by Patrick means that the objects are ordered in *Declination,* from 85° north to 80° south. No matter where you live, northern or southern hemisphere, you will have your own Caldwell objects to observe! The vast majority of the objects in Patrick's collection are eminently suitable for binocular observers.

If you are wondering why they should be named "Caldwell", the answer is that Patrick's surname is actually Caldwell-Moore. It would have been too confusing to use Moore as that would have created another set of "M" objects.

The number of Caldwell objects you can see from your own back-yard is immediately and easily calculated. In the northern hemisphere simply subtract (add in the southern hemisphere) 85 degrees from your latitude and this will give you your declination limit for objects that will rise at least 5° above your horizon. For those at say, 40° N in the USA

(Philadelphia, Indianapolis, Denver), that translates to 45°S. In the UK or Canada at 52°N, it translates to 33°S. The Caldwell object limit for this is C67. However, in these days of global travel there is no reason to be limited to your own location and observing the objects can be a continuing process of expanding your personal collection up to all the 109.

Appendix A gives the Catalogue in full and has visibility limits for several locations around the World. One fact that caught my eye is that a holiday in Barbados can reach from C1 to C108 – perhaps it would be worth checking out some holiday brochures!

The Caldwell Catalogue

The Caldwell Catalogue includes some spectacular objects. Even hardened deep sky observers, who may have thought they knew it all, will discover some new gems in Patrick's list – I have been observing for 40 years, and I did! They include a range of objects much wider than Messier's limited perspective (i.e. it might be confused for a comet), from the Hyades – the ultimate naked eye cluster – to possibly the most intriguing of all galaxies, Centaurus A. It even includes a dark nebula, the Coalsack! You could never confuse *that* for a comet.

The Caldwell Catalogue is a veritable cosmic zoo covering many types of astronomical objects. Along the way we will meet all manner of planetary nebulae, young and old open clusters, near and far globular clusters, nebulae (emission & reflection), galaxies of all shapes and sizes (edge-on/face-on/ peculiar), Wolf-Rayet stars, T Tauri stars, Blue Stragglers, X-ray objects, Fliers and LINERS. And of course lurking in the background are black holes!

Many of the objects are "Binocular Objects". That is not to say a telescope doesn't improve the view, rather that the catalogue is suitable for the humblest of optical aids. So don't think it is for telescope users only as nothing could be further from the truth. If binoculars can be held rock steady it is surprising what they can reveal – even resting them on a vehicle's roof improves the view and resolution. They are of course much easier to take to an out-of-town dark sky or on your holidays. Binoculars, away from urban sky-glow, can often show more than a telescope located on the edge of a city. Around 100 of the 109 objects can be seen with astronomical-type binoculars (20 × 80) although with the proviso that some of the smaller planetary nebula will be decidedly stellar in appearance but remain detectable. Some of the objects are actually better in binoculars than in telescopes.

For the visual observer, I have attempted to describe how each object looks in a range of telescope sizes. My own observations over that 40 year period and in my travels have been with 8 × 50 and 20 × 80 binoculars as well as telescopes of 6 inch, 8 inch, $12\frac{1}{2}$ inch and 16 inch (for the metric reader, that's 150 mm, 200 mm, 320 mm and 400 mm). Just to use my own observations would have been very restrictive and probably atypical – I'm not saying whether I am a worse or better observer than average! There had to be a better option. So I consulted as many observations by other observers as I could lay my hands on. They have been of differing skills (it's surprising what a difference skill makes!) and of course, using different apertures. What made this possible was the Internet, and specifically the World Wide Web. Here many many amateurs now publish what they have personally seen. They describe it in their own words and it is all the more refreshing for that. The equipment used has ranged from department store refractors to the biggest Dobsonian and the locations from urban driveways to ultra-dark mountain tops – a variety that no one person could ever hope to match. This extremely rich resource would have been impossible to draw upon only 10 years ago and it would have been a terrible waste not to make use of it today. Generally I have decided to concentrate on the more conservative observations and details, generally recorded by more than one person. For me, there is nothing worse than to read about fantastic detail being visible only to discover that it's otherwise when you get to the eyepiece. If you have exceptional eyesight or view from a very dark site then you should improve markedly on the views described here, which are those of a typical observer with average skills and from an averagely-good site.

I have laid the book out in as logical a format as I could devise. Each object has a full two-page spread. This should make photocopying (personal use only!) for taking out into the field, relatively easy. For each object we have a quality image, a comprehensive database, a finder chart, observational description for a variety of binocular/telescope sizes and, finally, the most up to date technical data.

One item that we debated was whether to include drawings. In the end I decided to omit them. The reason for this is that they often (at least to me) look very unconvincing. Deep sky observing is about chasing extremely subtle detail, which can often be only discerned with a lot of perseverance and experience. This is difficult to show in a drawing without exaggerating contrast and reality. How do you draw something that can only be seen with averted vision? So I have gone with a verbal description which should, in combination with the (generally CCD) image that does not suffer as much from burnout of bright details, enable a better appreciation of what is really there. In this way I hope the inexperienced observer won't be fooled into believing that all the detail can be seen in the eyepiece with just a cursory glance.

For each object therefore the following five topics are covered: Image, Database, Finder Chart, Visual Description and Object Description. The background and explanation for these follows.

Image

This is possibly the first deep-sky album to feature almost
exclusively CCD images by amateur astronomers. Only a
very few of the wide-field objects have been captured using
film. Whilst the occasional bleeding (vertical streaks) from
bright stars might, to the uninitiated, appear somewhat
unusual, the detectors and the images they produce have
many advantages over film. Their increased efficiency means
amateur telescopes (in skilful hands) can capture the faintest
of detail. The reduced burnout of bright details that I
mentioned earlier is another big advantage. For example,
deep photographs (on film) of globular clusters are
invariably totally lacking in detail. So it is only logical that
this modern collection features modern images, with over
100 of them captured using CCD cameras. Unless indicated
otherwise North is orientated to the top.

Collecting the images for all 109 objects proved to be a
daunting task. Whilst many astro-imagers or astro-
photographers are capable of producing a few stunning
images, only a dedicated few have a library of images
comprehensive enough – and of sufficiently high quality – to
cover all the comparatively less well known Caldwell Objects.
Even so, for several southern hemisphere objects I could find
no amateur image and I am extremely grateful to Steve Lee
for tracking down the "missing" objects and imaging them
especially for this book. As you will read in the description
for C100, Steve was also a big help in deciphering this
enigmatic object.

The imagers are, in alphabetic order:

Christian Buil & Eric Thouvenot – to CCD imagers these two
will need no introduction. Put simply they were the pioneers
of amateur CCD astronomy. Their Buil-Thouvenot (BT)
Atlas, produced in conjunction with several other talented
French amateurs, is the definitive source of high quality CCD
images. It contains nearly 5,000 images showing 5002
galaxies, 284 galaxy clusters, 645 open clusters, 136 globular
clusters, 155 planetary nebulae, 279 bright nebulae and 90
dark nebulae. To compile such a huge collection, as well as
their own Alcyone Observatory's Celestron C11, they had
access to the famous Pic du Midi Observatory's 60 cm
Newtonian and 1metre Cassegrain. CCD cameras utilised a
variety of detectors including an early Thomson (221 × 145
pixels), several based on the popular Kodak chips such as the
KAF400 (768 × 512 pixels) and the KAF1600 (1536 × 1024
pixels) and a large 512 × 512 pixels (19 micron pixel size)
Thomson with MPP (low noise) technology. Both the "*BT
Atlas*" (available from Sky Publishing on CD-ROM) and the
book explaining their techniques and methods, "*CCD
Astronomy*" by Christian Buil (published by Willmann-Bell),
are heartily recommended. Their images appear in this book
courtesy of Sky Publishing Corporation.

Steve Lee – I have already referred to Steve's efforts in
mopping up the elusive objects and he probably has every

Christian Buil and the T60 telescope at the Pic du Midi Observatory. It was here that many of the images for the BT-Atlas were taken.

amateur astronomer's dream job. He is a night assistant at the Anglo Australian Observatory in Australia, the home of David Malin. When he is not working (if you can call it working!), Steve is an amateur astronomer and also a skilful CCD imager. He is an accomplished telescope maker and, just for good measure, he also discovered a comet (Comet 1999 H1 Lee). His main telescopes are a 12 inch (306 mm) Newtonian for imaging work and a $12\frac{1}{4}$ inch (310 mm) Newtonian for his visual observations. His CCD camera is a home-made Cookbook type. If you are wondering how Steve achieves such high resolution with a Cookbook camera then the answer lies in the software he developed specially. This increases resolution when several images are combined.

Tim Puckett – this Georgian CCD imager probably has more published images to his name than any other amateur. When the CCD manufacturer *Apogee* runs an advertisement

Steve Lee with his 306 mm imaging Newtonian and Cookbook Camera. His observatory is located in Australia and both the telescope and the CCD camera are homemade.

Tim Puckett's 24 inch (600 mm) f/8 Ritchey-Chretien telescope. It was built by Tim and is located at his observatory in the North Georgia Mountains, USA.

for their cameras, they invariably use several of Tim's stunning colour images. His images are that good. Recently Tim has discovered a couple of supernovae in distant galaxies. It is difficult keeping up with his telescopes but currently on his home-built observatory complex are a 28 inch (700 mm) Cassegrain and a 24 inch (600 mm) Ritchey-Chretien. A Celestron C14 on a Paramount computerised mount is his latest addition. His workhorse 12 inch (300 mm) Schmidt-Cassegrain, which took some of the images used here, has now gone to pastures new. For Omega Centauri he used Alex Richter's 16 inch (400 mm) Schmidt-Cassegrain in South Africa during its initial setting up.

Alex Richter – this South African observer has perhaps every amateur astronomer's dream set-up with a computer controlled 16 inch (400 mm) Meade Schmidt-Cassegrain and an Apogee CCD camera. Whilst new to imaging, Alex had Tim Puckett help him get operational. He continued on his own to take several of the southern globular cluster images, after Tim had returned to the USA.

Pedro Re with his Celestron C14 located in Portugal

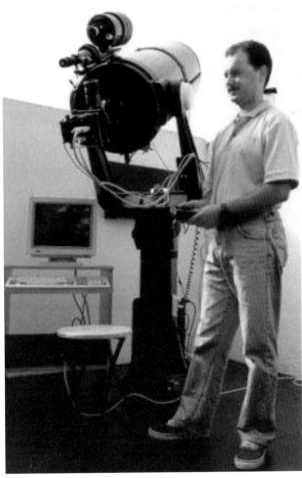

Pedro Re – this Portuguese amateur astronomer is a Marine Biologist by day and a prolific CCD imager by night. His images are widely published and after many years as a successful astro-photographer he began CCD imaging in 1994. His main telescopes are a Celestron C14 and a Meade 2045. On the CCD camera front he has an extremely wide range at his disposal, including the following types: HiSIS22, ST7, HX916 and Audine. He has published his own *CCD Deep-Sky Atlas* on CD-ROM, which encompasses all the Messier objects and many from the NGC/IC as well. Another very good buy.

Finally myself – after over 30 years taking astro-photographs I, like Pedro, began CCD imaging in 1994 when I purchased one of the first HiSIS CCD cameras in England. This featured the then new Kodak KAF400 chip with 768 × 512 pixels. After several years of learning the ropes I went on to edit and co-write the book *The Art and Science of CCD Astronomy* published by Springer. Telescopes used in my Lancashire Observatory for my Caldwell images were a 6 inch (150 mm) Newtonian, an 8 inch (200 mm) Schmidt Cassegrain and a 16 inch (400 mm) f/4.7 Newtonian. The latter telescope is the most recent addition and was built entirely by the Bolton Group of telescope makers of which I am a member. All three telescopes are mounted on the same fork mount, which is computer controlled.

David Ratledge's observatory in Lancashire, England housing his homemade 16 inch (400 mm) computer controlled telescope.

Database

The general database is derived from the Saguaro Astronomy Society's (SAC) well-known deep sky database, supplemented and corrected where necessary. This Astronomical Society has done an immense service by producing a database for amateur astronomers by amateur astronomers. They have corrected and updated much of the original *New General Catalogue* and the subsequent IC additions. The NGC, which was produced in the 19th Century by Dreyer, contained many anomalies and just plain errors. I have, however, included Dreyer's visual appearance (NGC Description), not in it its usual coded format but decoded into sensible English. These descriptions are extremely useful – but bear in mind Dreyer used large telescopes (up to 72 inches), so terms like "bright" or "resolvable" need putting into context. Cross-references to chart numbers in the popular star atlases have also been included, namely the *Millennium Star Atlas, Sky Atlas 2000* and *Uranometria*. These refer to the first editions of these atlases and should they subsequently be revised (as Uranometria is at the time of writing) they *may* not correspond to revised second editions. Other general source data has largely been based on the following:-

Open Clusters

The basic reference source for open clusters was the catalogue by Ruprecht, Balazs and White, 1981. This is available on the Internet from the Astronomical Data Centre (ADC) and its reference is 7101A. However, a much more comprehensive database of open cluster information is WEBDA. This Internet based site is devoted solely to open clusters and is run by Jean-Claude Mermilliod of the University of Lausanne, Switzerland. Virtually everything you could wish to know about each individual cluster is available here, from member star properties to an extensive bibliography.

The cluster type is that defined by Trumpler and is in 3 parts:

Concentration: I – detached, strong concentration towards the centre
II – detached, weak concentration towards the centre
III – no concentration towards the centre
IV – not well detached from surrounding field

Range in brightness: 1 – small range
2 – moderate range
3 – large range

Richness: p – poor (<50 stars)
m – moderately rich (>50, <100 stars)
r – rich (>100 stars)

Globular Clusters

There are over 150 globular clusters in our galaxy but from our viewpoint they are not evenly spread around the sky. Most are found in only one half of the sky and are concentrated towards the centre of our galaxy. In fact this is how, initially, it was realised where the centre of the Milky Way was located. They predominate in the southern hemisphere, especially in Sagittarius, Ophiuchus and Serpens. This means that northern summer is the best season in which to observe these ancient objects.

You will find the finest selection of globular clusters in the Caldwell Catalogue, far more so than the Messier catalogue. They range from the most spectacular to the most remote – something for everyone. If you ever get the chance to use a really big telescope, forget galaxies and take a look at a globular cluster. The myriad of stellar points is like nothing else and certainly better than any image or photograph can ever capture.

Globulars are very homogeneous objects and whilst superficially they may all appear the same they are not. The classification system for them is that devised by Lord Shapley in the 19th century. The classification is as follows:

1 – very concentrated
to
12 – not concentrated

The distances to globular clusters are taken from the Catalogue of Milky Way Galaxies by William E. Harris of McMaster University (May 1997) and are quoted in pc (parsecs). To convert parsecs to the more common light years, multiply by 3.26.

Bright Nebulae

"Bright nebulae" is a term that covers a mixture of types. They split into two main categories namely emission and reflection and only a short definition is given here. An emission nebula is a region of inter-stellar gas that shines because it is ionised by (usually) the ultra-violet radiation from hot young O and B stars. Supernova remnants are somewhat similar but more complex, for example the Cygnus Loop (C33/34) results from the collision of the nebula with the interstellar medium. This excites the gas and makes it visible. Reflection nebulae on the other hand shine purely by starlight reflecting off dust grains within them. The light is generally bluer than that of the illuminating star but the spectra is substantially the same. Some objects are a combination of the two types and sometimes even have a dark nebula as well!

Dark Nebulae

The Caldwell catalogue is unique in including a dark nebula (the Coalsack). Such nebulae are clouds of interstellar dust and gas that are sufficiently dense to obscure either partially or completely, the light from stars behind them. They are situated along the spiral arms and are only prominent where they mask the background Milky Way. They were first catalogued by E.E. Barnard (of the star fame) in 1919 and most are still known by their "B" number today although paradoxically "our" dark nebula, the Coalsack, does not have one!

Planetary Nebulae

The Caldwell collection includes a wide range of planetary nebulae, again far wider than the Messier catalogue. It includes the famous Cat's Eye Nebula plus the largest, the Helix Nebula, and the brightest, NGC 7293 (C63). The classification is according to Vorontsov-Velyaminov and is as follows:

1 stellar
2 smooth disc
   (a = brighter centre, b = uniform brightness, c = traces of ring structure)
3 irregular disc
   (a = very irregular brightness distribution, b = traces of ring structure)
4 ring structure

Galaxies

The source material for galaxies is derived from two catalogues. The first and best known is the 3rd *Reference Catalogue of Bright Galaxies* by de Vaucouleurs (1991/3). It is commonly known as RC3. This really is the definitive source. However, whilst it is extremely comprehensive, it lacks their distances. For these, the source data has been derived from the *Nearby Galaxies Catalogue* (1988) by Tully. This is commonly referred to as the NBG. Both catalogues are available on the Internet from the Astronomical Data Centre (ADC). Their references are 7155 and 7145 respectively. Now a "health" warning! Astronomical distances are notoriously subject to change and even in the last decade, as a result of the first findings from Hipparcos (the European Space Agency's astrometric and photometric space mission), distances to many objects have again been the subject of debate. So distances should be taken as an *indication* rather than as an absolute value.

The classification follows that derived by Hubble and modified by de Vaucouleurs in 1959. This is:

E – elliptical
    E0 – spherical
    to
    E7 – highly flattened
    subgroups: d – dwarf, c – supergiant, D – diffuse halo

S – spiral
    Sa – tightly wound arms with large nucleus
    to
    Sd – very loosely wound arms with tiny nucleus

SB – barred spiral
    subgroups as for S type

SAB – intermediate to S and SB with small bar
    subgroups as for S type

Pec – peculiar or distorted

Irr – irregular, no definite form.

d – dwarf

Finder Charts

I have in effect provided two finder charts. The first, the all-sky Caldwell chart, provides a quick reference of what is currently observable. This encompasses the whole sky by Mercator projection and unavoidably suffers from some distortion near the poles. It is however the most useful chart and you will no doubt return to it over and over again – I know I do.

The second is the individual chart with each object. These are limited to just the brighter naked-eye stars and constellation patterns. The exact object location is marked with "Telrad" circles. These popular 1x star finders have proved far better than conventional finders for locating faint deep-sky objects. They work by projecting 0.5°, 2° and 4° illuminated circles onto the night sky at the point at which the telescope is aiming. All that has to been done then is to move the telescope so that the rings appear as they do in the chart, and the object will be within a low-power field of view. They make locating even the most difficult object very easy and are highly recommended.

I have orientated the finder charts with due north vertical from the object – there is nothing more confusing than finder charts with north in all different directions. The scale of the charts varies to ensure sufficient bright stars being present to make location easy. The scale is of course easily determined by the Telrad circles – remember the largest is 4° across.

The charts were drawn using base maps derived from *Deep Space* by David Chandler. This DOS-based program is a little old fashioned in these "Windows" days but it is excellent for printed maps.

Generally, directions are given from the most convenient nearby bright star. Apart from those stars prominent enough to have their own well-known name, they are usually labelled with a Greek letter on the chart and by their English equivalent name in the description. The following Greek to English conversion table should therefore make it all straightforward, if you are not familiar with the Greek alphabet.

Greek	Name	Greek	Name	Greek	Name
α	Alpha	ι	Iota	ρ	Rho
β	Beta	κ	Kappa	σ	Sigma
γ	Gamma	λ	Lambda	τ	Tau
δ	Delta	μ	Mu	υ	Upsilon
ε	Epsilon	ν	Nu	ϕ	Phi
ζ	Zeta	ξ	Xi	χ	Chi
η	Eta	o	Omicron	ψ	Psi
θ	Theta	π	Pi	ω	Omega

Visual Description

As I said, I drew upon the extensive resource of the many observations published on the Internet. It is impractical to list all of these, but it would be wrong to omit the more prominent.

First amongst these is the IAAC, the *Internet Amateur Astronomers Calalog*(ue). This Internet forum is a repository for deep-sky observations by amateur astronomers around the world. It has a standard form (enter log) which is completed online and about a week later the observation will be posted on the IAAC web site. Don't think it is only for experts or big telescope owners – there are even naked-eye observations posted. The web site is run by Lew Gamer. Prominent amongst the IAAC observers are Lew Gamer himself, Steve Coe (of Saguaro Database fame) and Todd Gross (the Boston Meteorologist).

The second web site is MAC, the *Minimum Aperture Club*. This club's main aim is (as you might guess) to determine the smallest optical aid needed to see deep-sky objects. The site is run by Jere Kahanpaa and features many of his own observations. However, it is his excellent drawings of deep-sky objects which makes this web site unique. It is limited to declination 25° south.

The third is DOC, the *Deep-sky Observer's Companion*. Run by Auke Slotegraff, this web site is organised into objects visible for each chart of the *Uranometria* Sky Atlas. It too features many of the observations by its own webmaster (Auke Slotegraff) but also includes many observations by historic observers such as Sir John Herschel, Abbe Lacaille and James Dunlop. At the time of writing this, the web site only covers some of the southern hemisphere charts (from 17°S to the South pole) but when complete it should be a fabulous resource.

Object Description

It is frequently said that beauty is in the eye of the beholder. What often makes a deep-sky object interesting is knowing some intriguing background information about it. For many of the faintest objects, that's the *only* point of interest!

Even for brighter objects, that added information about what you are looking at definitely improves the view. For example, that central star in a planetary nebula might be an exotic Wolf-Rayet star or that smudge of a galaxy might have a giant black hole at its centre.

For many years *the* source of technical information on astronomical objects has been the three volumes of *Burnham's Celestial Handbooks*, published 1966–78. Unfortunately, time has not been too kind to them and today they provide more of an historical view. The sheer volume of modern research, particularly in the 1990s, has simply left them behind. So the source for technical information has had to be the latest research papers. For some of the more popular objects (with professional astronomers) I found nearly 1000 references in the technical publications, the vast majority of which were in the last twenty years. Clearly with the room available, we are after all dealing with 109 objects, I have had to be very selective and cherry-pick what I considered to be the most interesting or the very latest discoveries. The result is, I hope, a good cross-section of the current scope and extent of professional research taking place all over the World at the new millennium. Another modern change is that research is invariably now carried out by a large team. To save space only the first name is quoted with "et al." having to cover the many others involved.

Conclusion

Finally – when the guide book to the Pennine Way (a long-distance walk in Britain) was first published Wainwright, the author of the book, offered a pint of (warm) beer to all those who successfully completed the journey. Sorry, but I won't be emulating that particular offer!

However, I have included in the appendix a pair of blank certificates for those completing their own personal "Caldwell challenge", whether it be for all objects or just those visible from their location. By all means photocopy it and fill it in… I am sure you will have earned it and, more importantly, enjoyed yourself along the way.

The 109 Caldwell Objects

A Comprehensive Object-by-Object Guide

Image

David Ratledge

Database

Name and/or Catalogue Designations:
NGC 188
Collinder 6
Melotte 2

Type of object: Open Cluster

Catalogue position for epoch J2000.0
Right ascension: 00h 44m 24.0s
Declination: +85° 20' 00"

Constellation: Cepheus

Object information:
Magnitude: 8.1
Size: 14.0'
Number of stars: 150–200 to magnitude 18
Magnitude of brightest star: 12.1
Object classification: II 2 r
NGC Description: very large, round, 150 stars between magnitude 10 and 18
Note: very old open cluster

Star atlas chart numbers:
Millennium Star Atlas, Charts 5–6, Volume I
Sky Atlas 2000.0, Chart 1
Uranometria 2000, Chart 1, Volume 1

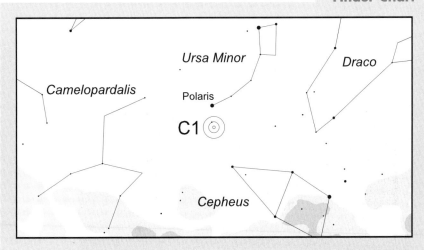

The NGC describes this open cluster as "very bright" but this should be taken with a pinch of salt. It is visible in an 8 inch (200 mm) telescope as just a circular soft glow. A larger telescope is required to resolve its stars as these are around 12th magnitude, although some stars (foreground?) are brighter. Its location, 4° south of Polaris towards Cepheus, makes it easy to locate but an equatorial telescope is cumbersome so close to the pole. So it is probably easier to locate it with an altazimuthly mounted telescope, such as a Dobsonian. Alternatively, if your telescope has an equatorial mount, set it up with the polar axis pointing away from the pole. Although not a stunning object it is well worth searching out because of its great age (see below).

The open cluster C1 (NGC 188) has been the subject of numerous studies since Allan Sandage (1962) emphasised its importance as one of the oldest open clusters. Its richness and location, relatively far away from the galactic plane, gave the first indications that C1 could well be as old as the galactic disc itself and therefore important in understanding its formation. However, whilst no longer the oldest known cluster, much recent research (Twarog 1989, Demarque et al. 1992, Maynet et al. 1993, Dinescu et al. 1995) has put its age at around 6 billion years. It has an interesting population of stars, which includes a large number of contact variables (with their possible later evolution into FK Comae objects), several faint blue straggler stars (stars bluer than they should be for their age) and a surprising star (known as II-91), which is a possible sub-dwarf. A recent study by Von Hippel/ Sarajedini (1998) concentrated on white dwarfs in C1 and their work found a possible error in earlier photometry. This indicated that C1 was certainly older than 1.14 billion years and they pointed out that their data does not rule out the suggested age of 6 billion years.

Tim Puckett

Database

Name and/or Catalogue Designations:
NGC 40
H IV 58
PK 120+ 9.1

Type of object: Planetary Nebula

Catalogue position for epoch J2000.0
Right ascension: 00h 13m 02.4s
Declination: +72° 31′ 40″

Constellation: Cepheus

Object information:
Magnitude: 10.7
Size: 1.0′ × 0.7′
Magnitude of central star: 11.5
Object classification: 3b(3)
NGC Description: faint, very small, round, very suddenly much brighter towards the middle, large star in contact following
Note: planetary with Wolf-Rayet central star

Star atlas chart numbers:
Millennium Star Atlas, Charts 23–24, Volume I
Sky Atlas 2000.0, Chart 1
Uranometria 2000, Chart 3, Volume 1

Finder Chart

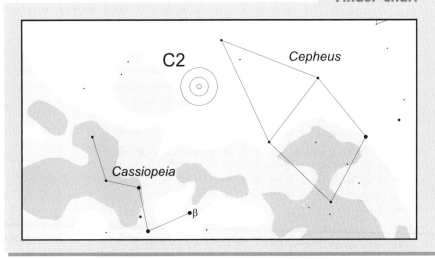

Visual Description

An excellent planetary nebula, well worth searching out as it displays more detail than most. It appears stellar at low power with the nebulosity revealing itself at higher magnification in just an 8 inch (200 mm) telescope. The central star is reasonably bright at about magnitude 11.6. There are reports of a UHC filter helping to make the halo of nebulosity much more conspicuous. In any case with higher powers and larger apertures (over 12 inches) significant detail appears with a notable brightening towards the edge of the disc. Lord Rosse even thought he saw spiral structure. It is not easy to find however, with no nearby bright stars – try going 13° north from Beta Cassiopeiae towards Polaris.

Object Description

C2 is a low-excitation planetary nebula (Aller & Czyzak, 1979) and is powered by a central Wolf-Rayet star of type WC8 (Louise, 1981 and Benvenuti et al., 1981). Wolf-Rayet stars are extremely hot and have erratic mass ejections plus, in this case, it is possibly a binary. Planetary nebula are thought to form when an ageing star ejects its outer material, the remaining exposed hot stellar core then excites the nebula causing in to glow (fluoresce). A commissioning image of C2 taken with the 3.5 metre WIYN telescope on Kitt Peak has helped explain much about this object. The Wolf-Rayet star has a mass of around 0.7 solar masses, and is much hotter (around 90,000°K) than would be expected just from the properties of the surrounding nebula. This suggests the presence of shielding material between the star and the glowing nebula. Such higher density material could form in the shock interface between the fast wind from the central star (about 1800 km/s) and the nebular shells themselves. It is thought the nebular material would cover about 25% of the sky as viewed by the central star, implying rather asymmetric mass-loss from the star, now in its giant branch stage. C2 has an extended halo, not normally visible, probably caused by earlier mass ejections.

Image

Buil-Thouvenot

Database

Name and/or Catalogue Designations:
NGC 4236
H V 51
UGC 7306
IRAS12140+6947
PGC 39346

Type of object: Galaxy

Catalogue position for epoch J2000.0
Right ascension: 12h 16m 43.4s
Declination: +69° 27' 56"

Constellation: Draco

Object information:
Magnitude: 9.6
Size: 23.0' × 8.0'
Position angle: 162°
Object classification: SB(s)dmIV
NGC Description: very faint, extremely large, much extended P.A. 160°, very gradually brighter to the middle
Notes: 8th mag. Star 13.45' north of nucleus. Part of the M81 group

Star atlas chart numbers:
Millennium Star Atlas, Charts 535–536, Volume II
Sky Atlas 2000.0, Chart 2
Uranometria 2000, Chart 25, Volume 1

Finder Chart

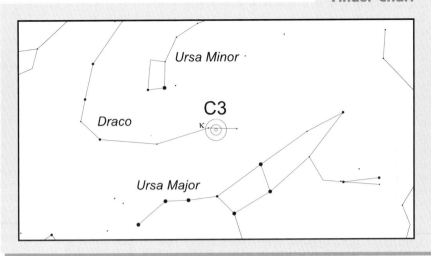

Visual Description

I first "discovered" this galaxy when searching galaxy listings for nice big ones at northerly declinations, suitable for my site in northern England. C3 (NGC4236) certainly is big (18 arc-minutes long) but it is virtually unknown. Why? The simple answer is that it is much fainter than its integrated magnitude of 9.6 indicates. It is however well worth searching out and under dark skies its long central bar is quite clear, even in an 8 inch (200 mm) telescope. The bar is orientated NNW – SSE. Our angle of view means the spiral arms (see below) are not seen at their best. Even in a 16 inch (400 mm) I could not detect any indication of them. C3 is located in a pretty barren part of Draco with Kappa Draconis the closest guide star. This 3.9th magnitude star is around 12° north of the Plough. When you've located it, head 1.5° WSW.

Object Description

C3 is a classical barred spiral with a very prominent bar (hence SB) and fainter spiral arms. It is not unlike C44. The difference is our angle of view. The bar in C3 is approximately at right angles to our view and is therefore seen to advantage. On the other hand, the plane of the spiral arms is nearly edge-on to us and consequently they are close to the bar and only just distinguishable from it. The term "bar" was created by Hubble in his classic 1926 paper on galaxy morphology. This class had previously been known as Phi-type after its similarity to Greek letter Phi (ϕ). The commonness of spiral galaxies with bars indicates that they can form them without much difficulty and retain them for significant periods of time. The distance to C3 has been the subject of an investigation by Tikhonov et al. (1991). This team used the giant 6 metre Russian telescope to determine the colour magnitude diagram for C3. Based on the properties of the observed red giants they place C3 at a distance of just over 10 million light years, which agrees with an earlier calculation based on Cepheid variable stars by Freedman and Madore (1988).

C4

Image

Buil-Thouvenot

Database

Name and/or Catalogue Designations:
NGC 7023
H IV 74

Type of object: Reflection Nebula

Catalogue position for epoch J2000.0
Right ascension: 21h 01m 12.0s
Declination: +68° 10′ 00″

Constellation: Cepheus

Object information:
Magnitude: 7.1
Size: 13.0′
Object classification: reflection nebula
NGC Description: 7th mag. star in extremely faint nebula, extremely large
Note: central star HD200775

Star atlas chart numbers:
Millennium Star Atlas, Charts 1061–1062, Volume III
Sky Atlas 2000.0, Chart 3
Uranometria 2000, Chart 33, Volume 1

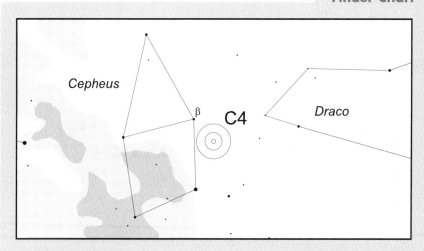

Visual Description

One of the easier reflection nebulae to see – certainly easier than that of the better known one around the Pleiades. C4 is fairly small but what helps its visibility is contrast. Immediately surrounding it is dark material, probably un-illuminated parts of the same nebula, which blocks any background stars, making the nebula more prominent. The central star, at 7th magnitude, is easy in all telescopes and with an 8 inch (200 mm) a round glow about 10 to 15 arc-minutes in diameter is evident, albeit a bit featureless. However you might be able to make out a brighter knot just south of the central star. To locate C4 it is probably easiest to star-hop from T Cephei. This star, although variable between 6th and 10th magnitude, is easy to spot because of its strong red colour even when at its faintest. T Cephei is 3° south-west from Beta Cephei. In the same field as T Cephei will be the 7th magnitude 67 Cephei. Then C4 is just over half a degree west of 67 Cephei.

Object Description

C4 is classed as a bright reflection nebula and is illuminated by HD200775, which is a Herbig Ae/Be star. The full spectrum of this star was determined by Sellgren et al. (1998) and the "e" in Ae/Be denotes that at least some of its hydrogen lines present are in *emission* rather than the more normal absorption. The ultra violet radiation from this star is thought to have photo-dissociated the molecular hydrogen of the nebula into atomic hydrogen – the star is not hot enough to form an H-II region. Initially however, outflow from the star (around 5,000 years ago) cleared a circum-stellar cocoon, which exposed the material to the UV radiation. Most of the atomic gas in our galaxy is believed to have been processed in such a manner i.e. outflow followed by photo-dissociation. There is an excellent set of lecture notes on C4 published by Chris Rogers on the Internet. The URL is:-

http://zebu.uoregon.edu/crogers/talk/conclusions.htm

Image

David Ratledge

Database

Names and/or Catalogue Designations
IC 342
UGC 2847
IRAS03419+6756
PGC 13826

Type of object: Galaxy

Catalogue position for epoch J2000.0
Right ascension: 03h 46m 49.7s
Declination: +68° 05′ 45″

Constellation: Camelopardalis

Object information:
Magnitude: 9.2 (12.0 nucleus)
Size: 17.8′ × 17.4′
Object classification: S(B)c
NGC Description: faint, very large, round, very small bright nucleus
Note: heavily obscured by the Milky Way

Star atlas chart numbers:
Millennium Star Atlas, Charts 31–32, Volume I
Sky Atlas 2000.0, Chart 1
Uranometria 2000, Chart 18, Volume 1

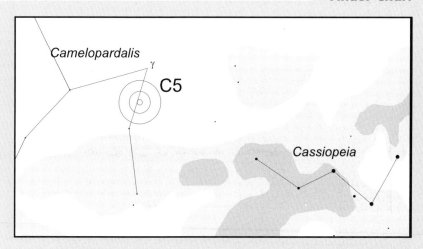

Visual Description

Don't be fooled by its 9th magnitude listing, a much more accurate guide to its visibility is its 12th magnitude nucleus. The fact that it is an IC catalogue object means it was missed by all the Herschels, despite being an object that would have been high in the sky from their observing site in England. In fact it is so faint that it was not discovered until 1890 by W.F. Denning and it wasn't until 1934 that Hubble and Humason, using the Mount Wilson telescope, determined it was a galaxy. A fact that an 8 inch telescope and a CCD camera, as our image shows, can easily verify today. There are reports of it being visible in an 8 inch (200 mm) telescope but a very dark sky and a 16 inch is required for a reasonable view of this large faint object. Even so do not expect to see spiral arms, just a circular faint haze, larger than you might imagine. It lies 3.2° south of Gamma Camelopardalis (mag. 4.59).

Object Description

This face-on barred spiral galaxy lies only 10° above the plane of Milky Way and as a result it is heavily obscured. Nevertheless its extent has been traced out to over 30 arc-minutes giving an apparent size around double that catalogued for it. C5 is one of the nearest galaxies outside our local group but its actual distance is somewhat disputed. A distance of 6 million light years was determined independently by both McCall (1989) and Madore & Freedman (1992) but Krismer et al. (1995) have suggested a somewhat larger distance of 10 million light years. It is classed as a "starburst galaxy" as it is undergoing massive star formation. The central region of C5 has been studied in the near-infra-red by Boker et al. (1997). They found complex dynamical processes dominated by a cluster of young massive super-giants at its very centre. Surrounding this was a ring of molecular gas with two intense star forming H-II regions on either side of the nucleus. These they compare to 30 Doradus (C103) in the Large Magellanic Cloud. Their conclusion is that we are currently witnessing a "bar-driven central starburst".

Planetary Nebula in Draco

Image

Pedro Re

Database

Name and/or Catalogue Designations:
Cat's Eye Nebula
NGC 6543
H IV 37
PK 96+29.1

Type of object: Planetary Nebula

Catalogue position for epoch J2000.0
Right ascension: 17h 58m 35.1s
Declination: +66° 37′ 53″

Constellation: Draco

Object information:
Magnitude: 8.3
Size: 22.0″ × 16.0″
Magnitude of central star: 11.14
Object classification: 3a(2)
NGC Description: very bright, pretty small, suddenly brighter towards the middle with very
    small nucleus
Note: central star HD164963

Star atlas chart numbers:
Millennium Star Atlas, Charts 1065–1066, Volume III
Sky Atlas 2000.0, Chart 3
Uranometria 2000, Chart 30, Volume 1

Finder Chart

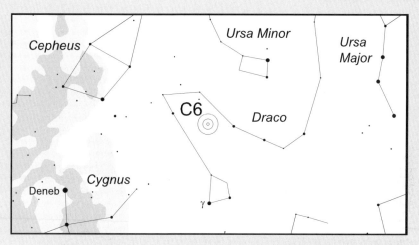

Visual Description

Since the publication of the stunning images of the Cat's Eye Nebula taken by the Hubble Space Telescope (HST), this object has become one of the best known of all planetary nebulae. Despite requiring the HST to reveal all its glory it is easily spotted in modest amateur telescopes at low power. A 6 inch (150 mm) is sufficient to show a bright oval with a definite blue/green tinge to it (unlike the false colour HST image, which shows it red with green extremities). To see the central star a much larger telescope and higher magnification is required – generally a 16 inch (400 mm) is needed – although central brightening is evident in a 10 inch (250 mm). To find it, to it is probably easiest to first locate Gamma Draco and then move 15° north in declination towards Polaris.

Object Description

The famous HST images showed, in high resolution, one of the most complex and beautiful planetary nebulae ever seen. These images (credit: Harrington & Borkowski, 1995) revealed incredible intricate structures including concentric gas shells, jets of high-speed gas and unusual shock-induced knots of gas. The simple interpretation is for a fast stellar wind of material, blown off the central star, which interacts with earlier shells. However, to explain its complex morphology it is likely that the nebula has been produced by a binary star system, with the stars too close together for even Hubble to separate. The interaction of the two stars producing the complex object. Many of C6's peculiar features had been detected earlier (Balick & Preston, 1987) where a double bipolar nebula was proposed as an explanation but such a structure is difficult to explain with a single central star. More recently, Zweigle et al. (1997) using far ultra-violet observations taken from on board the Space Shuttle (known as ORFEUS), determined the properties of C6's central star, which is generally classed as a Wolf-Rayet. This they found to have a mass 0.6× that of the Sun, a luminosity 3000× the Sun and a surface temperature of 50,000°K.

Image

David Ratledge

Database

Names and/or Catalogue Designations
NGC 2403
H V 44
UGC3918
IRAS07321+6543
PGC 21396

Type of object: Galaxy

Catalogue position for epoch J2000.0
Right ascension: 07h 36m 54.5s
Declination: +65° 35′ 58″

Constellation: Camelopardalis

Object information:
Magnitude: 8.5
Size: 17.8′ × 10′
Position angle: 127°
Object classification: Sc
NGC Description: considerably bright, extremely large, very much extended, very gradually
    much brighter middle nucleus
Note: possible member of M81/2 group

Star atlas chart numbers:
Millennium Star Atlas, Charts 25–26, Volume I
Sky Atlas 2000.0, Chart 1
Uranometria 2000, Chart 21, Volume 1

Finder Chart

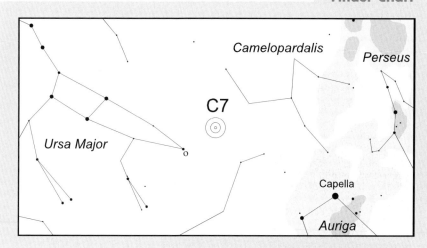

Visual Description

This object is one of the best-kept secrets in the northern hemisphere and deserves to be much better known. C7 was discovered by William Herschel and it is easily visible today in a modern 6 inch (150 mm) telescope. Under reasonably dark skies even binoculars will reveal it. Observed from Britain, it rises high overhead making it easy to see, even through light polluted suburban skies. It becomes distinctly oval in an 8 inch (200 mm) telescope and with averted vision the brighter western spiral arm is hinted at. In a 16 inch (400 mm) telescope the western and northern spiral arms are the clearest. It is not easy to find with Camelopardalis having no significant stars so it is best to star-hop from Omicron Ursae Majoris and head 7.5° north-west.

Object Description

This Sc type spiral galaxy is viewed not quite face-on and is similar in structure to M33 in Triangulum. Like M33 it has a small bright nucleus and many emission nebula in it – over 100 have been identified. It also has many brilliant blue giant stars populating its spiral arms. It is however outside the local group and is probably an outlying member of the M81/2 group in nearby Ursa Major. It was in fact the first galaxy outside the local group to have identified in it, Cepheid variable stars, by Tamman & Sandage (1968). These are one of the most important of "standard candles" used for calculating distances. However determining periods and luminosity for these variable stars at this distance is not a trivial task. Their data has been re-assessed by Madore (1976), Hanes (1982) and Rowan-Robinson (1984) with no general agreement! But more recently the Hubble Space Telescope, as part of its project to determine the Hubble Constant (Freedman), has now measured several Cepheids in C7 to new levels of accuracy. The current estimate of the distance to C7 is 11.5 million light years.

Open Cluster in Cassiopeia

David Ratledge

Name and/or Catalogue Designations:
NGC 559
H VII 48
Collinder 13

Type of object: Open Cluster

Catalogue position for epoch J2000.0
Right ascension: 01h 29m 30.0s
Declination: +63° 18' 00"

Constellation: Cassiopeia

Object information:
Magnitude: 9.5
Size: 4.4'
Number of stars: 60
Magnitude of brightest star: 10.8 (TYC 4035-913-1)
Object classification: II 2 m
NGC Description: cluster, bright, pretty large, pretty rich
Notes: very old cluster

Star atlas chart numbers:
Millennium Star Atlas, Charts 33–34, Volume I
Sky Atlas 2000.0, Chart 1
Uranometria 2000, Chart 16, Volume 1

Finder Chart

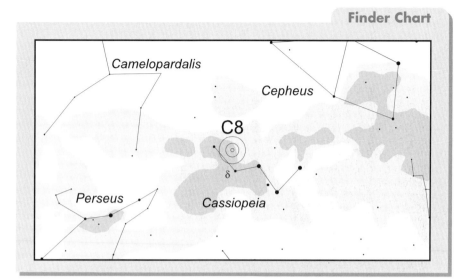

Visual Description

As C8 is not one of Cassiopeia's more spectacular open clusters it has largely been passed by, that is until now! This small cluster will however take some teasing out of the general star field. Not easy to spot with binoculars, it really demands astronomical ones (20×80s) to see it for certain and even then it is just a faint haze. To get amongst its member stars an 8 inch (200 mm) telescope is required and do not be fooled by that catalogue listing of 60 stars in the cluster. At this aperture expect to see only half a dozen or so. It really does need more aperture and a 16 inch (400 mm) if you want to be impressed. The problem is its stars are faint (that Milky Way obscuration doesn't help – see below). In that 16 inch it gives the impression of splitting into two parts, a brighter north-east and a fainter south-west. It is located north-west of the first two stars (Epsilon & Delta) of the W of Cassiopeia – about 3° north of Delta Cassiopeiæ (mag. 2.7).

Object Description

C8 is a very old open cluster, approximately 1.3 billion years in age, located about 400 light years away. It is partially obscured by the Milky Way suffering a 0.7 magnitude decrease (reddening) of its stars. It did have its moment of glory with the discovery in its midst of the radio source G 127.3+0.7. Caswell and Pauls (1977) identified this as a supernova remnant. If the supernova remnant was physically associated with the cluster then an improved estimate of the distance to the cluster might make the supernova remnant a useful distance calibrator. However Pauls et al (1982) and Milne (1988) have shown that it is not associated with the cluster and is just a line of sight coincidence. Xilouris et al. (1992) were the first to optically record the supernova remnant, previously it had only been detected using radio telescopes. It is estimated to be 18,000 years old and lie beyond C8.

Bright Nebula in Cepheus

Buil-Thouvenot

Name and/or Catalogue Designations:
Cave nebula
Sh2-155

Type of object: Bright Nebula

Catalogue position for epoch J2000.0
Right ascension: 22h 56m 48.0s
Declination: +62° 37′ 00″

Constellation: Cepheus

Object information:
Size: 50.0′ × 30.0′
Object classification: emission
NGC Description: none
Note: No. 155 in "*A Catalogue of H-II Regions*" by Sharpless, 1959

Star atlas chart numbers:
Millennium Star Atlas, Charts 1069–1070, Volume III
Sky Atlas 2000.0, Chart 3
Uranometria 2000, Chart 34, Volume 1

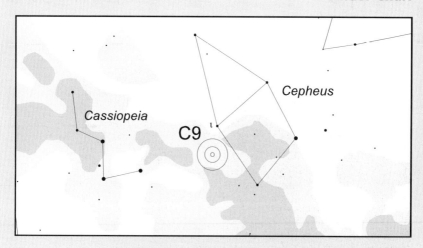

Visual Description

C9 is a very photogenic and an extremely interesting emission (H-II) nebula but visually it represents quite a challenge. Although believed to have been named by M. Wolf, it was first catalogued by Sharpless in his "Catalogue of H-II Regions". His catalogue was complied from objects identified on the Palomar Observatory Sky Survey (POSS) photographs taken with the 48 inch (1.2 metre) Schmidt Camera. C9 is close to the 6.54 magnitude star HD216945 and to have any chance of seeing C9 this star has to be found. The brightest part of the nebula is a crescent shape (concave to the east) and is located 12 arc-minutes to the north-west of HD216945. There is an 8.76 magnitude star (HD217061) involved in it and a slightly brighter star (HD 217086) 6 arc-minutes further north. Having got the precise location it is now time to slip in that UHC nebula filter because without it there is no hope! A huge aperture is not required but your lowest power is. The Minimum Aperture Club (MAC) lists a 5 inch (125 mm) as the minimum. Just expect to see the faintest of amorphous glows. To get to HD216945 head 3.75° SSE of Iota Cephei (mag. 3.5).

Object Description

C9 is just part of one of the most complex and studied areas of the sky, namely the Cepheus B/Sh2-155/Cepheus OB3 Association/Cepheus Molecular Cloud complex! Cepheus B is the hottest part of the giant Cepheus molecular cloud where star formation is taking place today. Nearby is the Cepheus OB3 association. The visible interface between the molecular cloud and the OB stars is the part we know as the Cave Nebula or C9. The sharp edges of C9 clearly indicate the presence of an ionisation front. The brightest members of the OB3 (HD217061/217086) are thought to be largely responsible for the UV radiation that is ionising the nebula. Very recently (1999) Naylor & Fabian reported the discovery of over 50 X-ray point sources within the complex, the majority of which they interpreted as T Tauri stars (variable proto-stars).

C10

Open Cluster in Cassiopeia

David Ratledge

Name and/or Catalogue Designations:
NGC 663
H VI 31
Collinder 20
Melotte 11

Type of object: Open Cluster

Catalogue position for epoch J2000.0
Right ascension: 01h 46m 00.0s
Declination: +61° 15′ 00″

Constellation: Cassiopeia

Object information:
Magnitude: 7.1
Size: 16.0′
Number of stars: 80
Magnitude of brightest star: 8.4
Object classification: III 2 m
NGC Description: cluster, bright, large, extremely rich, star preceding large (bright)
Notes: in Cassiopeia OB8 association, includes double stars Struve 151, 152 & 153

Star atlas chart numbers:
Millennium Star Atlas, Charts 47–48, Volume I
Sky Atlas 2000.0, Chart 1
Uranometria 2000, Chart 16, Volume 1

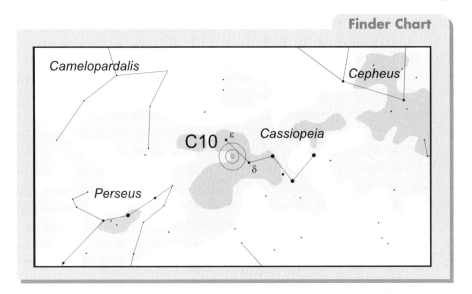

Visual Description

Cassiopeia has many fine clusters and C10 is one of its best but surprisingly little known. A dark sky and 10 × 50 binoculars are all that is needed to appreciate this sparkling large cluster. I first "discovered" it in just such binoculars 30 years ago when searching for the nearby M103 but stumbled on this much more impressive cluster instead. To start resolving its many stars a 6 inch (150 mm) telescope is perfect. Use a low power as the cluster appears larger than its 16 arc-minute catalogue size – I estimate around 30 arc-minutes in its longer axis of north-east to south-west. In that 6 inch about 40 stars are visible. Upping the telescope aperture to an 8 inch, and with more magnification, then some of the brighter stars are seen to be doubles with the total number of stars passing 50 quite easily. It is simple by find by heading 1° east, at right-angles from an imaginary line joining Epsilon to Delta Cassiopeiae. There are several clusters in the vicinity but C10 is the biggest.

Object Description

C10 is a very young open cluster and 1991 was a big year for it as two separate research groups produced data for it. Firstly Phelps & Janes determined the luminosity function for C10 and noted a significant deficit of low mass stars. The same year, Tapia et al., also studied the photometry of this cluster in the near infra-red, visible and ultra-violet parts of the spectrum. They were able to determine that C10 lies at a distance of about 8,000 light years and has an age of around 9 million years, although more recent research puts its age a touch younger at 4 million years. However C10's main claim to fame is that it has the highest percentage of Be stars known in any open cluster. Be stars are B-type stars with the unusual property of having some of the H lines in *emission* (hence the e) rather than the normal absorption (see C97 for more information on Be stars). One of its Be stars appears to be a high mass X-ray binary (Motch et al., 1991).

C11

Image

Tim Puckett

Database

Name and/or Catalogue Designations:
Bubble Nebula
NGC 7635
H IV 52

Type of object: Bright Nebula

Catalogue position for epoch J2000.0
Right ascension: 23h 20m 42.0s
Declination: +61° 11′ 00″

Constellation: Cassiopeia

Object information:
Magnitude: 8.69 (central star)
Size: 15.0′ × 8.0′
Object classification: emission
NGC Description: very faint, 8th magnitude star involved, little eccentric
Notes: Near M52

Star atlas chart numbers:
Millennium Star Atlas, Charts 1069–1070, Volume III
Sky Atlas 2000.0, Chart 3
Uranometria 2000, Chart 34, Volume 1

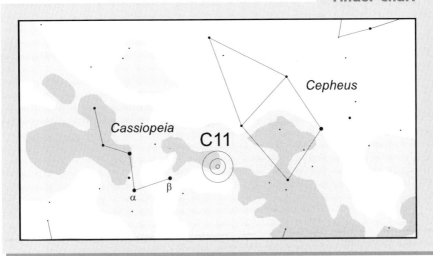

Visual Description

Easy to star-hop to, easy to find but not easy to see! The open cluster M52 points the way, C11 is just one degree away in a south-westerly direction. Observing from my suburban skies with an 8 inch, 12.5 inch and 16 inch, only in the latter was the nebula seen. All telescopes could see the 9th magnitude star BD+6002522 and show it a bit "fuzzy", but only the 16 inch could see the round glow of the Bubble for definite. Increasing the magnification showed some detail but the bubble shape was not evident. In out of town skies an 8 inch should see traces of it and, as the nebula has a strong O-III emission, an O-III nebula filter should make observation easier. To find M52, and hence C11, follow the line Alpha to Beta Cassiopeiae and on in a straight line for the same distance.

Object Description

Billed by the Hubble Heritage Team (Hester et al.1992), as the "bubble versus the cloud". The part we know as the Bubble Nebula is being forced out by a stellar wind of ionised gas from a massive central star. However next door is a giant molecular cloud (a portion of which is visible in our image). The cloud, although able to contain the expansion of the bubble's gas, gets blasted with intense radiation from the central star. This radiation heats up denser regions, ionising it and causing the shell to glow. This interpretation was confirmed by Buckalew et al. (1999) who, using observations with the Space Telescope Imaging Spectroscope (STIS), concluded that the characteristic rim of C11 was the edge of a shell of ionised gas that is being "snowploughed" through the surrounding region by the supersonic wind from the star. C11 is actually the smallest part of three bubbles surrounding the central star and part of a gigantic network known as S162, created by other massive stars. C11's central star, BD+602522, is 10 to 20 times more massive than the Sun and is classed as type O6.5iiif. The nebula is about 10 light years across and 11,300 light years away.

C12

Image

Pedro Re

Database

Name and/or Catalogue Designations:
NGC 6946
H IV 76
UGC 11597
ARP 29
PGC 65001

Type of object: Galaxy

Catalogue position for epoch J2000.0
Right ascension: 20h 34m 52.0s
Declination: +60° 09′ 15″

Constellation: Cepheus (partially in Cygnus)

Object information:
Magnitude: 8.8
Size: 14.0′
Object classification: SAB(rs)cd
NGC Description: very faint, very large, very small bright towards the middle, partially
    resolved some stars
Note: formerly classified as Sc

Star atlas chart numbers:
Millennium Star Atlas, Charts 1073–1074, Volume III
Sky Atlas 2000.0, Chart 3
Uranometria 2000, Chart 32, Volume 1

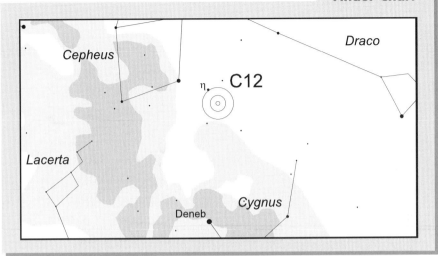

Visual Description

Although this face-on galaxy is one of the nearest galaxies to us (beyond the Local Group that is) it is a difficult object visually. This is because its position in our sky is close to the plane of the Milky Way, which results in it being partially obscured and much reduced in contrast by foreground stars. The nucleus is also small and faint but represents the best chance to spot it. Certainly from my suburban skies with a 12.5 inch (32 cm) telescope that was all I could detect, although there are sightings of an outer haze with an 8 inch (200 mm) from darker locations. To see the five main spiral arms, a large Dobsonian and pristine skies are required. Despite its faintness it is surprisingly easy to locate its position. This is because the open cluster NGC6939 is nearby. This cluster is a splendid sight in any size telescope and a low power field can include both. C12 is 1° south-east of the cluster and 2° SW of Eta Cephei (mag. 3.4).

Object Description

C12 was for many years classed as an Sc galaxy i.e. similar to M33 with loose open arms and small nucleus. However it has recently been categorised as an SAB(rs)cd galaxy. This classification refers to the presence of a small core with multiple well-defined arms (cd), a poorly-developed bar across the middle (AB) and an inner confused ring (rs). It is undergoing a starburst phase in its nuclear region. Engelbracht et al. (1996) concluded that much of the starburst is masked by dust and that it has had a duration of around 20 million years. Elmegreen et al. (1998), observing in the infra-red, found evidence for a bar like structure that could be fuelling the starburst. There have been several supernovae observed in C12 and it also has the brightest X-ray supernova remnant (SNR) known anywhere. Blair et al. (1997) postulated colliding SNRs but Dunne et al. (1999) preferred a supernova explosion inside a dense interstellar bubble as an explanation. C12 is thought to be about 18 million light years away.

C13

Open Cluster in Cassiopeia

Image

David Ratledge

Database

Name and/or Catalogue Designations:
The Owl Cluster or Phi Cas Cluster
NGC 457
H VII 42
Collinder 12
Melotte 7

Type of object: Open Cluster

Catalogue position for epoch J2000.0
Right ascension: 01h 19m 06.0s
Declination: +58° 20′ 00″

Constellation: Cassiopeia

Object information:
Magnitude: 6.4
Size: 13.0′
Number of stars: 80
Magnitude of brightest star: 8.6
Object classification: I 3 r
NGC Description: cluster, bright, large, pretty round irregular, stars 7,8,10 magnitude
Note: 0.5 degrees north-west is the cluster NGC436

Star atlas chart numbers:
Millennium Star Atlas, Charts 47–48, Volume I
Sky Atlas 2000.0, Chart 1
Uranometria 2000, Chart 36, Volume 1

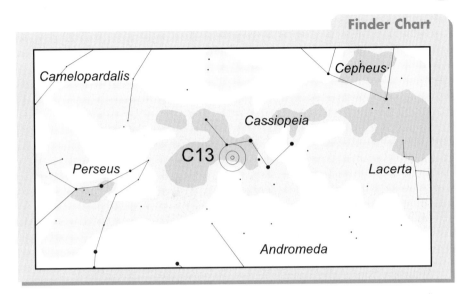

Visual Description

There are two open clusters in Cassiopeia with Messier numbers (M52 and M103) but, in my opinion, C13 is by far the brightest and prettiest cluster in this constellation. What make it so spectacular is the presence of two bright, probably foreground stars (Phi Cassiopeiae and HD7902), and a plethora of sparkling cluster stars, at least 80, in a 12.5 inch (320 mm) telescope. Not that a large telescope is required, binoculars give a fine view. An 8 inch (200 mm) will resolve around 40 stars and show the curving rows of 10th magnitude stars that have lead to it being called, the Owl Cluster. However its likeness to an owl requires a vivid imagination and its name is more probably the result of its appearance in photographs where the two bright stars, mentioned above, appear as a pair of prominent "eyes" with the cluster forming the owl's body. Because of the presence of the 5th magnitude star Phi Cas in the cluster its location is straight forward, 2° below the "W" of Cassiopeia.

Object Description

The stellar population of C13 is similar to that of the young Double Cluster in Perseus (C14), although on a reduced scale. According to Eggen (1982), C13 is slightly older than those two clusters. The membership of the bright stars, Phi Cassiopeiae and HD7092 (the owl's eyes) is a point of discussion, particularly as the former is classed as a type F0 1a (a supergiant). Their distances have been determined by Hipparcos (Tycho Catalogue) and found to be 2330 +/-762 and 2760 +/-1188 light years respectively. The distance to the cluster proper is harder to establish but Fitzsimmons (1992) used photometric methods to calculate a distance of 2.5 kpc, which equates to just over 8,000 light years. Even taking the upper range of the Hipparcos values the probability is, therefore, that they are foreground stars. C13 is located in the Perseus spiral arm and Dufton et al. (1994) used the 2.5 m Isaac Newton Telescope to determine the chemical composition of B-type stars in C13 finding a surprising variation to other clusters in this spiral arm.

C14

Image

David Ratledge

Database

Name and/or Catalogue Designations:
Double Cluster In Perseus or the Sword Handle

"h" Persei	"chi" Persei
NGC 869	NGC 884
H VI 33	H VI 34
Collinder 24	Collinder 25
Melotte13	Melotte 14

Type of object: Open Cluster

Catalogue position for epoch J2000.0
Right ascension: 02h 19m 00.0s
Declination: +57° 09' 00"

Constellation: Perseus

Object information:
Magnitude: 4.3
Size: 30.0'
Number of stars: 200
Magnitude of brightest star: 6.6
Object classification: I 3 r
NGC Description: Cluster, very very large, very round irregular, stars 7...14
Note: both at a distance of 7,000 light years

Type of object: Open Cluster

Catalogue position for epoch J2000.0
Right ascension: 02h 22m 24.0s
Declination: +57° 07' 00"

Constellation: Perseus

Object information:
Magnitude: 4.4
Size: 30.0'
Number of stars: 150
Magnitude of brightest star: 8.1
Object classification: I 3 r
NGC Description: Cluster, very large, very round irregular, ruby star in middle

Star atlas chart numbers:
Millennium Star Atlas, Charts 45–46, Volume I
Sky Atlas 2000.0, Chart 1
Uranometria 2000, Chart 37, Volume 1

Finder Chart

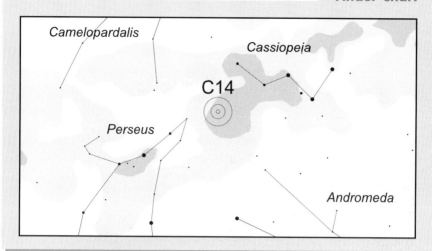

Visual Description

Plainly visible to the naked eye just north of the body of Perseus, the Double Cluster was first catalogued by the Greek Hipparchus (145BC). They are one of the showpieces of the northern hemisphere and are at their best in binoculars or a small telescope. Here the field of view can be wide enough to encompass both clusters and, at the same time, show a little bit of the Milky Way around them. This sets them off a treat. A wide-angle eyepiece, giving at least a 1° true field, is required to achieve this. An unusual feature of the clusters is the presence of red super-giant stars, a prominent one between them and several in NGC884. NGC869 is the more compact with a central clump of stars. My best ever view of them was from the dark skies of the Lake District with 8 inch (200 mm) binoculars. They produced a breathtaking 3D effect – the clusters appearing as sparkling points, hanging in front of the Milky Way.

Object Description

Although open clusters are common in the Milky Way, these two are exceptional because they are so rich in young (O and B) stars and are so close together. Their association is not just a line of sight coincidence, they both share a common ancestry (Pismis, 1959) being part of the much larger Perseus OB1 Association. At most they are only a few hundred light years apart (tiny on a galactic scale) and of a similar, very young age. Until recently, NGC 884 was thought to be the younger at 3.2 million years old and NGC869 slightly older at 5.6 million years and the closer of the two. However, Waelkens et al. (1990) using photometry of the two clusters over an 8 year period have cast some doubt on this, suggesting that previously claimed age and distance differences were probably spurious. This group also found that at least half the stars in these clusters were variable with most being of the Be type – very young stars with an emission feature in their spectrum thought to be indicative of a circumstellar disc. Many of the stars also had high rotational velocities, which makes interpretation of them more difficult.

C15

Image

Buil-Thouvenot

Database

Name and/or Catalogue Designations:
Blinking Nebula
NGC 6826
H IV 73
PK 83+12.1

Type of object: Planetary Nebula

Catalogue position for epoch J2000.0
Right ascension: 19h 44m 51.0s
Declination: +50° 31′ 20″

Constellation: Cygnus

Object information:
Magnitude: 8.8
Size: 27.0″ × 24.0″
Magnitude of central star: 11.0
Object classification: 3a(2)
NGC Description: planetary, bright, pretty large, round, central star 11th magnitude
Note: has faint outer halo 2.3 arc-minutes diameter

Star atlas chart numbers:
Millennium Star Atlas, Charts 1109–1110, Volume III
Sky Atlas 2000.0, Chart 3
Uranometria 2000, Chart 55, Volume 1

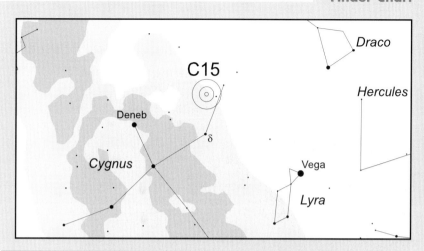

Visual Description

C15, the Blinking Nebula, is one of the strangest objects in the northern sky. The blinking effect is wholly in the eye of the beholder and is not seen by everyone. The generally accepted view as to why the nebula (appears) to blink is to do with the way our eyes work when looking at faint objects. It would seem that the colour (wavelength) of the nebula itself is a good match to the wavelength at which the rods in our eyes are most receptive. So when observing the nebula with averted vision i.e. its light is falling on the rods in our eyes, it is bright and prominent. However when observing directly, using the cones in our eyes, it all but disappears. The result is that it appears to blink in and out of view as we move our eye around. The central 11th magnitude star conversely becomes more or less conspicuous as the nebula does the opposite. The effect is visible in all sizes of telescope from 6 inches (150 mm) to 20 inches (500 mm). It is not an easy object to locate being in a particularly rich area of the Milky Way. With an equatorial telescope is easiest to find by first locating Delta Cygni and moving 6° due north.

Object Description

This is a well studied planetary nebula (PN) and Balick (1987) showed that C15 contains 2 distinct regions: a bright elliptical inner region, about 26 arc-seconds in diameter, and a faint spherical halo, about 2.3 arc-minutes in diameter. The inner region he found to be quite complex and included a bright elliptical rim, a bright shell and two ansae just outside the rim. The halo is somewhat filamentary and it also has a bright shell at its outer edge. The general model (Balick) for PN is for the hot central star to emit a fast wind, which catches up with the slow wind emitted previously in the star's red giant stage, but this does not fully explain the complexities observed here. Plait & Soker (1990) examined various scenarios and were able to explain a spherical outer shell and an inner elliptical inner region by the presence of a low mass companion to the progenitor star. This companion did not survive and was either evaporated or collided with the central star.

Open Cluster in Lacerta

Image

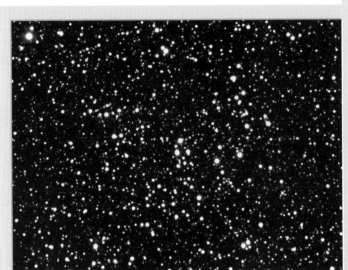

David Ratledge

Database

Name and/or Catalogue Designations:
NGC 7243
H VIII 75
Collinder 448
Melotte 240

Type of object: Open Cluster

Catalogue position for epoch J2000.0
Right ascension: 22h 15m 18.0s
Declination: +49° 53' 00"

Constellation: Lacerta

Object information:
Magnitude: 6.4
Size: 21.0'
Number of stars: 40
Magnitude of brightest star: 8.5
Object classification: IV 2 p
NGC Description: cluster, large, irregular compressed, stars very large (bright)
Note: listed as one of the (USA) Astronomical League's Urban Objects and one of the Texas Star Party's binocular objects.

Star atlas chart numbers:
Millennium Star Atlas, Charts 1103–1104, Volume III
Sky Atlas 2000.0, Chart 9
Uranometria 2000, Chart 57, Volume 1

Finder Chart

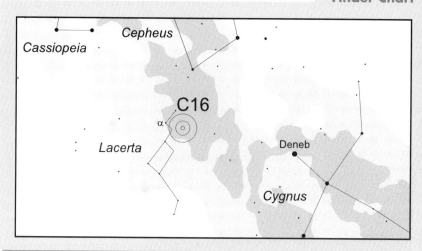

Visual Description

The constellation of Lacerta is a bit of an enigma. Although the Milky Way runs through its northern half it is much overlooked and not noted for its deep sky gems. It does however have three excellent open clusters, all binocular objects, and the middle one of which is C16 (NGC7243). It is visible in finderscopes as its brightest stars are between magnitudes 8 and 9 and to my eyes are perhaps a touch bluish. The cluster is large, about the size of the full moon. Its fainter stars go down to magnitude 13, so an 8 inch (200 mm) telescope is best but keep to a low power – try for a 1° field to set it off nicely. An unusual feature of this cluster is the uneven distribution of its 50 or so stars. It is almost as if it were two clusters having a divide running north-west to south east. There is a nice double star near the middle of the cluster (magnitude 9 with a separation of about 5 arc-seconds). To locate C16 it is easiest to star-hop from Alpha Lacertae (mag. 3.75) and then head 2.5° west. In a low power eyepiece the cluster should be at the southern edge of your field of view. Those other two clusters in Lacerta? They are NGC 7243 (north of C16) and NGC 7209 (south).

Object Description

The distance to C16 is currently estimated to be 2400 light years. Hipparcos has however measured the distances to several of the stars within the field of this cluster. These are between 25 and 500 light years distant so must be foreground objects, not to be unexpected with C16 being so close to the plane of the Milky Way. A classical way of determining cluster membership has been radial (line of sight) velocity measurements using spectroscopy. Radial velocity values for 29 stars in this cluster were determined by Hill & Barnes (1971). Another method is proper motion studies but these usually require measurements over a long period to detect (sideways) motion. However, Jilinski et al. have used plates taken in 1897 and 1996 at the Pulkovo Observatory. They were able to measure the proper motion for no less than 2593 stars!

C17

Image

Tim Puckett

Database

Name and/or Catalogue Designations:
NGC 147
UGC 326
PGC 2004

Type of object: Galaxy

Catalogue position for epoch J2000.0
Right ascension: 00h 33m 11.6s
Declination: +48° 30′ 28″

Constellation: Cassiopeia

Object information:
Magnitude: 9.5
Size: 13.0′ × 8.0′
Position angle: 25°
Object classification: dE4
NGC Description: very faint, very large, irregular round, gradually small bright middle, star mag. 1.1 (?)
Notes: 1° west of C18 (NGC185)

Star atlas chart numbers:
Millennium Star Atlas, Charts 85–86, Volume I
Sky Atlas 2000.0, Chart 4
Uranometria 2000, Chart 60, Volume 1

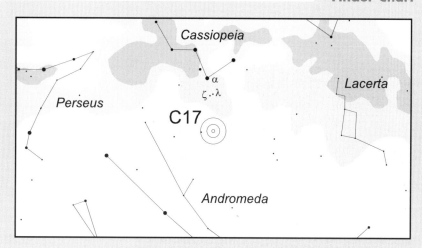

C17 is quite faint and tricky to see. Forget the integrated magnitude of 9.3, it is far too spread out for this to be a true representation of its brightness. In an 8 inch (200 mm) telescope, even when observing from dark skies, it is likely that averted vision will be needed just to detect it. Doubling the aperture (and magnification) makes it far easier and shows a small nucleus with at least one star overlaying the galaxy's hazy outline. Its size is surprisingly large, almost a quarter of a degree long. To locate it, try star-hopping from Alpha Cas to Lambda Cas and then head exactly 6° south towards Andromeda.

Usually classed as a dwarf elliptical galaxy (sometimes spheroidal), it is a companion of the Andromeda Galaxy (M31), the dominant galaxy of our Local Group. The Local Group was first recognised by Hubble in the early days of extra-galactic research when distances to galaxies were first being measured. We now recognise over 30 members of this condensation. Most of the light (and mass) in the Local Group is contained in the two large spiral galaxies, the Milky Way and M31, with the majority of the known smaller members being satellite galaxies of these two. The mass of M31 can be estimated from the orbit of C17. This has been calculated by Saha et al. (1990), who used observations of RR Lyrae variables to determine C17's position relative to M31. That distance was 500,000 light years which, together with C17's line of sight velocity, meant that M31's mass was at least 720 billion solar masses. On the other hand, the team of Han et al. (1996) used HST Wide Field/Planetary Camera 2 (WFPC2) to study the stellar populations in C17. Unlike M31's other companions (C18 and NGC 205), C17 appears to have no dust clouds to form new stars. Han et al. found that stellar formation ceased at 1 billion years ago with the younger stars more centrally concentrated than the older ones.

C18

Image

Tim Puckett

Database

Name and/or Catalogue Designations:
NGC 185
H II 707
UGC 396
IRAS00362+4803
PGC 2329

Type of object: Galaxy

Catalogue position for epoch J2000.0
Right ascension: 00h 38m 58.0s
Declination: +48° 20' 18"

Constellation: Cassiopeia

Object information:
Magnitude: 9.2
Size: 12.0' × 10.0'
Position angle: 35°
Object classification: dE0
NGC Description: pretty bright, very large, irregular round, very gradually much brighter
    towards the middle, mottled – not well resolved
Notes: 1° east of C17 (NGC 147)

Star atlas chart numbers:
Millennium Star Atlas, Charts 85–86, Volume I
Sky Atlas 2000.0, Chart 4
Uranometria 2000, Chart 60, Volume 1

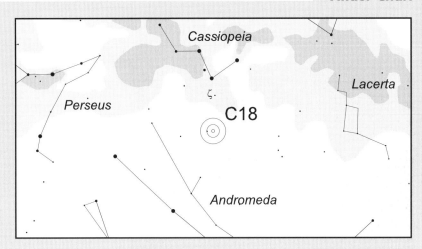

Visual Description

C18, the other of the pair of northerly satellites of M31, is the easier one to see and is quite distinct in an 8 inch (200 mm) telescope. I have seen it with my original 6 inch telescope many years ago (when skies were better) and, with averted vision, it is even visible in 4 inch refractors. It is surprising therefore that Herschel missed it although it is a low surface brightness (LSB) object. In larger telescopes (16 inch) it becomes a big elliptical glow $(14' \times 12')$ but largely featureless apart from a brighter core, which becomes more pronounced as magnification increases. The core though is not as prominent as in C17. The dark patch near the core was not detectable in my 16 inch. Despite being a satellite of the great Andromeda Galaxy, it is easiest to star-hop to it from Cassiopeia, being 5.4° south of Zeta Cassiopeiae (mag. 3.7).

Object Description

C18 is classed a dwarf elliptical galaxy (sometimes classed as dwarf spheroidal) and, like C17, is a satellite of M31 and therefore a member of the Local Group of Galaxies. Its stars were first resolved by Baade in 1944 and recent photometric work using the Hubble Space Telescope (STIS instrumentation), studied stars down to 27th magnitude (Tolstoy, 1998). The latter found a strong red clump/horizontal branch population of stars, which provided a better understanding of the old stellar population, in this and similar dwarf galaxies. For many years C18 has been a curiosity as its centre contains both optical dust clouds and luminous blue stars. The latter, according to Freedman et al. (1993), are the remains of a burst of star formation 20 to 40 million years ago. An interesting recent study concerning C18 and its relationship or otherwise with C17 is that by Van de Bergh (1998). Contrary to a previous claim, he found evidence that this pair form a stable binary system. Distance estimates place this pair on the near side of the M31 group and the fact that they still exist today is taken as suggesting that they do not have a plunging orbit and therefore have remained relatively undisturbed by M31.

C19

Image

Tim Puckett

Database

Name and/or Catalogue Designations:
Cocoon Nebula
IC 5146 (cluster)
Sh2-125 (nebula)

Type of object: Bright Nebula with Cluster

Catalogue position for epoch J2000.0
Right ascension: 21h 53m 24.0s
Declination: +47° 14' 00"

Constellation: Cygnus

Object information:
Magnitude: 10.0
Size: 9.0'
Number of stars: 20
Magnitude of brightest star: 9.6
Object classification: IV 2 p n
NGC Description: faint, large, irregular round, stars 9.5 mag. involved in bright and dark
    nebula
Notes: Cocoon nebula is at the end of the dark streamer, B168

Star atlas chart numbers:
Millennium Star Atlas, Charts 1103–1104, Volume III
Sky Atlas 2000.0, Chart 9
Uranometria 2000, Chart 86, Volume 1

Finder Chart

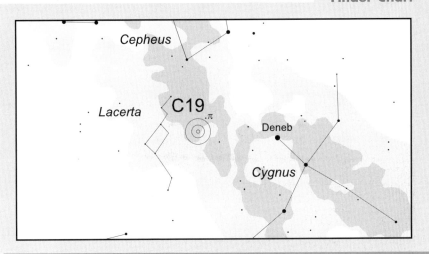

Visual Description

For those northern observers attempting to see all the Caldwell objects to within 5 degrees of their horizon, this one represents a tricky hurdle to overcome. It is extremely faint even in big telescopes. However there is an escape! C19 is a complex object with several components, namely a cluster (IC5146), a bright nebula (Sh2–125) and a dark nebula (B168). The term bright nebula is relative so don't get your hopes up but the other two are relatively easy. The cluster is not very rich with only around 20 stars visible but the brightest, bang in the middle, is magnitude 9.6 so it is visible even in a small telescope. The dark nebula is surprisingly a binocular object and, from dark skies, is quite noticeable as an area without stars. That leaves us with the nebula itself and the only chance is from a dark sky with a narrow band nebula filter (e.g. O-III) plus a medium sized telescope. It is located towards the north-east corner of Cygnus but with no convenient bright guide stars. The best chance is that Pi1 and Pi2 Cygni point towards it. All three are roughly in a line and equally spaced, with C19 to the south of them.

Object Description

The Nebula itself is both a reflection and emission type with its own dust lanes. The complex is classed as a molecular cloud with a young star cluster embedded within it. The cluster is dominated by BD+46°3474, a B0V star, which excites the symmetrical H-II nebula, Sharpless 125 (the Cocoon). In the model proposed by Roger & Irwin (1982), the Cocoon's ionised gas (H-II) has broken through the surrounding dark cloud while the bulk of nebular remains behind BD+46°3474, obscuring background stars. H-alpha emission-line stars were discovered in C19 by Herbig (1960). Forte et al. (1984) estimated the stellar population to be around 110 stars most of them were probably contracting objects with an age 1 to 3 million years and spectral class K. However, Lada et al. (1999) using infrared images were able to penetrate much deeper into the nebula and were able to find five times as many heavily obscured proto-stars.

Bright Nebula in Cygnus

Image

David Ratledge

Database

Name and/or Catalogue Designations:
North American Nebula
NGC 7000
H V 37?

Type of object: Bright Nebula

Catalogue position for epoch J2000.0
Right ascension: 20h 58m 30.0s
Declination: +44° 33′ 00″

Constellation: Cygnus

Object information:
Magnitude: 6.0
Size: 175.0′ × 110.0′
Object classification: emission
NGC Description: faint, extremely extended large, diffuse nebulosity
Note: part of the Cygnus X region and includes the cluster NGC 6997

Star atlas chart numbers:
Millennium Star Atlas, Charts 1125–1126, Volume III
Sky Atlas 2000.0, Chart 9
Uranometria 2000, Chart 85, Volume 1

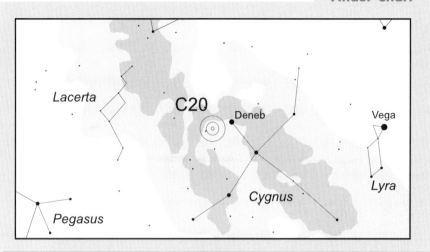

Visual Description

Although thought to have first been observed by William Herschel, it was a photograph taken by Dr Max Wolf in 1890, which resulted in it receiving its common name, the North American Nebula. Deneb points the way to this nebula but C20's low surface brightness can make it a difficult object to track down. It is very much an out of town dark sky object where it even becomes visible to the naked eye. The trick is to look for the dark "Gulf of Mexico" and the "Atlantic Coast" areas. In a telescope it all but disappears, although nebula filters can reveal its edges where these are distinct and contrasty, such as around the "Gulf of Mexico" referred to above. However 10 × 50 or larger binoculars probably provide the best view of this 2 degree wide object. Photographically it records easily and a simple telescope drive with piggyback camera can produce spectacular results.

Object Description

The American Nebula and the Pelican Nebula to its west are one and the same H-II nebula. They are the outlying areas of a complex known at the Cygnus X region with the central brightest part masked from our view by an intervening large dark cloud. It is this dark cloud that gives both these nebulae their shape. Radio telescopes have been able to penetrate the dark cloud and map its extent proving it to one object. Originally it was thought that Deneb was the source of illumination for this nebula but its ultra-violet emissions appear too weak. The star HD199579 has been suggested in the past but Wendker et al. (1983) thought this was a background star and found evidence for multiple excitation centres rather than one individual star. They observed cavities in the nebula containing very young stars, a region dominated by outflows from the cavities and further out regions ionised by leaking radiation from the stars. A follow up paper (Heske & Wendker, 1995) confirmed this scenario and Chavarria et al. later observed the young stars in C20 finding pre-main sequence objects of the types known as Herbig emission stars (Ae/Be) and T Tauri.

C21

Image

Pedro Re

Database

Name and/or Catalogue Designations:
NGC 4449
H I 213
UGC 7592
PGC 40973

Type of object: Galaxy

Catalogue position for epoch J2000.0
Right ascension: 12h 28m 11.4s
Declination: +44° 05' 40"

Constellation: Canes Venatici

Object information:
Magnitude: 9.6
Size: 6.0' × 4.5'
Position angle: 45°
Object classification: Irregular (Ibm/Sm)
NGC Description: very bright, cluster(?), much extended, double or bifid, well resolved, faint stars, bright centre with nucleus or star
Notes: C21 is a galaxy not a cluster as NGC classification

Star atlas chart numbers:
Millennium Star Atlas, Charts 611–612, Volume II
Sky Atlas 2000.0, Chart 7
Uranometria 2000, Chart 75, Volume 1

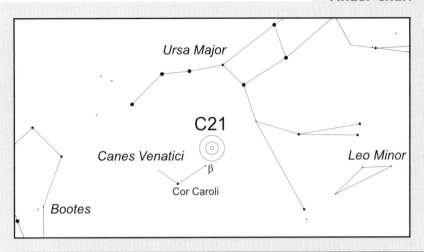

Visual Description

Caldwell 21 could well be mistaken for a comet (but not by readers of this book of course) as its profile is decidedly fan shaped. It is a relatively nearby galaxy and as such it is resolvable in large telescopes with even small ones able to show at least one H-II region (see below). An 8 inch (200 mm) telescope will display a wedge shaped haze with the sharp end WSW and the blunt end ENE. Under good skies, also visible is a small core. The best chance to see an H-II region in it is the knot in the north corner of the blunt end. In a 16 inch (400 mm) additional knots are visible including ones either side of the nucleus, quite close in. Visually, this is a surprisingly rewarding galaxy. It is located 3° NNW of Chara, Beta Canum Venaticorum (mag. 4.2).

Object Description

This dwarf irregular galaxy is a member of the Canes Venaticorum group of galaxies and is at a distance of about 4 Mpc (13 million light years). Its morphology, absolute size and brightness are remarkably similar to our own satellite galaxy, the Large Magellanic Cloud (LMC), even with traces of a bar, which contains older stars. In the northern part of the galaxy are many H-II regions (red) with ongoing star formation within them. Analysis of Hubble WFPC images by Seitzer & Grebel, (1998) showed a population of several hundred extended objects centred on the galaxy, which they interpreted as populous star clusters with a brightness similar to those in the LMC. Ultra-violet observations by Home et al. (1995) found many "starburst knots" and the presence of hot young stars with ages less than 12 million years. X-ray observations by Della Ceca et al. (1997) deduced that the source of "hard" X-rays was young binaries and/or supernova remnants and that for "soft" X-rays was due to hot diffuse gas. Radio observations (Bajaja et al. & Hill et al., both 1994) showed that NGC 4449 is embedded in a huge gaseous halo with a diameter of 14 times the optical one.

Planetary Nebula in Andromeda

Pedro Re

Name and/or Catalogue Designations:
Blue Snowball Nebula
NGC 7662
H IV 18
PK106-17.1

Type of object: Planetary Nebula

Catalogue position for epoch J2000.0
Right ascension: 23h 25m 54.3s
Declination: +42° 32′ 30″

Constellation: Andromeda

Object information:
Magnitude: 8.6
Size: 17.0″ × 14.0″
Magnitude of central star: 13.2
Object classification: 4(3)
NGC Description: magnificent planetary or annular nebula, very bright, pretty small, round, blue
Notes: HST found "fliers"

Star atlas chart numbers:
Millennium Star Atlas, Charts 1119–1120, Volume III
Sky Atlas 2000.0, Chart 9
Uranometria 2000, Chart 88, Volume 1

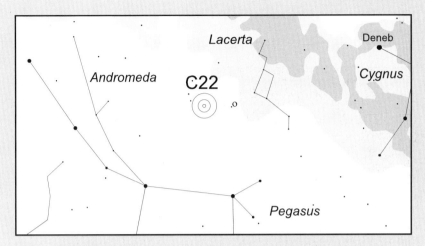

Observing Autumn planetary nebulae usually means M57 (The Ring) and M27 (The Dumbbell). Yet there is a third one in the same vicinity and equally bright, equally spectacular, the Blue Snowball. But it is virtually ignored. The Caldwell catalogue puts that right! With a visible magnitude of 8.6, C22 is even detectable in large binoculars. In a modest telescope (8 inch–200 mm) it comes into its own as a nearly circular blue patch, about the size of Saturn's disc. Just as one might expect a blue snowball to look like! In a 16 inch (400 mm) telescope the colour is stunning (aquamarine – blue with a touch of green) and a ring structure is evident when the magnification is increased. The ring is subtle and nothing like as prominent as that of M57. However it is the colour that will stay long in your mind. It is located about 14° above the Square of Pegasus with Omicron Andromedae (mag. 3.6) on the same declination as C22 and 4.4° to the west of it.

It is estimated that 95% of all stars that we see in our own galaxy will ultimately become planetary nebulae. This includes the Sun. The process, by which a star in its red-giant stage ejects its outer layers leaving a very hot dwarf at its centre to excite the nebula, is reasonably well understood. Or so it was thought. Images from the Hubble Space Telescope (Bruce Balick et al. 1993) have revealed some mystery objects, dubbed fliers, inside several planetary nebulae, including C22. Fliers (Fast Low-Ionisation Emission Regions) are knots of dense gas, which appear to have been ejected from the central star *before* it cast off the planetary nebula. Yet their shapes revealed in the Hubble images seem to suggest that they are stationary, and that material ejected from the star flows past them. Currently, no model can account for their formation or longevity. In 1996 Lame & Pogge again examined C22 finding several anomalies in that the nebula appeared to be older and the star less massive than theoretical predictions.

C23

Galaxy in Andromeda

Image

David Ratledge

Database

Name and/or Catalogue Designations:
NGC 891
H V 19
UGC 1831
IRAS02193+4207
PGC 9031

Type of object: Galaxy

Catalogue position for epoch J2000.0
Right ascension: 02h 22m 33.1s
Declination: +42° 20′ 48″

Constellation: Andromeda

Object information:
Magnitude: 9.9
Size: 14.0′ × 3.0′
Position angle: 22°
Object classification: Sb
NGC Description: bright, very large, very much elongated on position angle 22°
Note: inclination 1.4° from edge-on (Rupen, 1991)

Star atlas chart numbers:
Millennium Star Atlas, Charts 101–102, Volume I
Sky Atlas 2000.0, Chart 4
Uranometria 2000, Chart 62, Volume 1

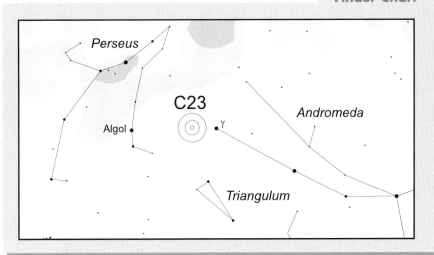

One of Autumn's gems and, in my opinion, the finest of the edge-on galaxies. C23 is extremely easy to find being on the same declination as Gamma Andromeda and approximately 3.5 degrees to the east of this star. Alternatively it can be found by "star hopping" from the open cluster M34, which is a similar distance away on its other side i.e. east. In an 8 inch (200 mm) telescope its elongated spindle shape is obvious and lies roughly east-west. The dark dust lane is much harder to spot – it is in fact not dark at all, just less bright! Increased magnification and averted vision helps but in a 16 inch (40 cm) telescope it becomes more obvious. Astronomical 20 × 80 binoculars, under good conditions (held steady, at a dark site and using averted vision), will render the slender shape of C23 visible.

C23 is the classical edge-on spiral galaxy and even though its spiral arms are invisible from our viewpoint, its prominent dust lane and central bulge makes its spiral classification certain. It is thought to be part of a group that includes nearby galaxies NGC1023 and 925. The current estimate of its distance is 31 million light years. C23 has been extensively studied over most of the electromagnetic spectrum and is noted for extensive "extraplanar gas" i.e. gas far from its galactic plane. Recent high quality images taken with the large 3.5 metre WIYN telescope on Kitt Peak (Howk & Savage, 1997) revealed a network of hundreds of dust clouds and streamers above and below this galaxy's plane. How this matter is propelled out of the disc is the puzzle. Some of the dust streamers observed suggested a supernova driven fountain or chimney phenomena thought to be tracing the violent disc-halo interface. However other dust clouds suggested a lower energy origin. Alton et al. (1998), using the newly commissioned Submillimetre Common-User Bolometer Array, produced the deepest images yet of any galaxy in this waveband. They confirmed the presence of dust chimneys extending over 6000 light years.

Galaxy in Perseus

Image

NGC1277
IC1907
NGC1278
NGC1274
NGC1273
NGC1275
NGC1272
NGC1270

Pedro Re

Database

Name and/or Catalogue Designations:
Perseus A
NGC 1275
3C84
UGC 2669
PGC 12429

Type of object: Galaxy

Catalogue position for epoch J2000.0
Right ascension: 03h 19m 48.5s
Declination: +41° 30′ 45″

Constellation: Perseus

Object information:
Magnitude: 11.9
Size: 3.5′ × 2.5′
Position angle: 110°
Object classification: peculiar Seyfert galaxy
NGC Description: faint, small
Notes: centre of Perseus Cluster, distance c. 300 million light years

Star atlas chart numbers:
Millennium Star Atlas, Charts 97–98, Volume I
Sky Atlas 2000.0, Chart 4
Uranometria 2000, Chart 63, Volume 1

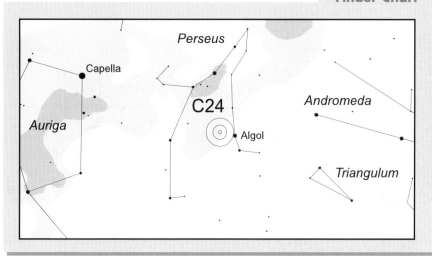

Visual Description

This galaxy has an appearance that belies its importance. At 12th magnitude C24 is not a binocular object and even an 8 inch (200 mm) telescope will need good skies just to glimpse it. Generally a 12 to 16 inch is required for a clear view and telescopes of this size will also show some of its fainter companions. Of these, NGC1272 (west of C24) and 1277/8 (north) are probably the least difficult but big apertures and dark skies will show many many more. To locate C24 sweep 2° east of Algol (Beta Persei) and then about 30 arc-minutes north.

Object Description

This object, which dominates the Perseus Galaxy Cluster, has over the years been classified as an elliptical, peculiar, blazar and currently a peculiar Seyfert galaxy. Its appearance, in images taken with the Hale 200 inch, is of a spherical galaxy swallowing a spiral. Its name derives from its strong radio emissions and it is also a prodigious emitter of X-rays. C24 is the dominant super giant galaxy at the heart of the Perseus Cluster of Galaxies (Abell 426) which, according to Egikian et al. (1985), has at least 530 members. This cluster is in turn part of the Pisces-Perseus (PP) supercluster. The latter is one of two great nearby superclusters – the other, located on the opposite side of the sky, is the infamous "Great Attractor". The PP supercluster takes the form a prominent extragalactic ridge from Pegasus through the Perseus cluster and on eastwards, disappearing behind the Milky Way. Research by Holtzman (1992), using early Hubble Space Telescope images, discovered that C24 is surrounded by at least 50 bright blue objects believed to be massive globular clusters. Their strong blue colour, which is common to them all, is a bit of a problem as this colour implies youth. The explanation offered is that they were most likely formed when two galaxies collided thus strengthening the view that this galaxy is in fact a merged object. Later HST images by Carlson et al. (1998) extended the number of compact clusters in C24 to a staggering 3,000. They found evidence for two distinct age groups, again strong evidence for a merger.

C25

Image

David Ratledge

Database

Name and/or Catalogue Designations:
NGC 2419
H I 218

Type of object: Globular Cluster

Catalogue position for epoch J2000.0:
Right ascension: 07h 38m 08.5s
Declination: +38° 52′ 55″

Constellation: Lynx

Object information:
Magnitude: 10.4
Size: 4.1′
Magnitude of Brightest Stars: 17
Object classification: 2
NGC Description: pretty bright, pretty large, little extended PA 90 degrees, very gradually
    brighter to the middle, star 7 to 8 magnitude PA 267 degrees and 4′ distant
Note: very distant cluster.
Distance from the Sun: 82.3 kpc.
Distance from the galactic centre: 89.6 kpc.

Star atlas chart numbers:
Millennium Star Atlas, Charts 107–108, Volume I
Sky Atlas 2000.0, Chart 5
Uranometria 2000, Chart 69, Volume 1

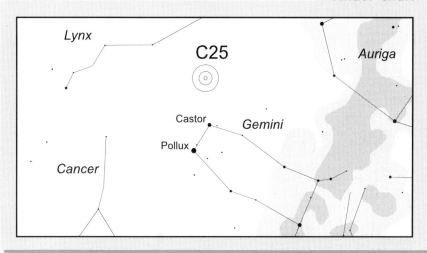

Visual Description

William Herschel called it a bright nebula and he was right as it is easily visible in a modern 8 inch (200 mm) telescope. However do not expect to resolve this globular's faint haze. The best that a large telescope (16 inch) can expect to reveal is a mottled appearance – the stars are simply too faint to see individually. C25 is very easy to locate being approximately 7° north of Castor and forms an equally spaced triple with two brightish (7–8th magnitude) stars on its western side, which point the way to it. Astronomical binoculars show these stars and, with averted vision, a hazy patch next to them.

Object Description

It was Lord Rosse with his giant 72 inch telescope in Parsonstown, Ireland who first suspected that this object might be a globular cluster. Its brightest stars are only magnitude 17, so to be the first to resolve it the largest telescope of its day was required. Nevertheless it is one of the most interesting of the Milky Way's globular clusters being the most remote known. Its distance at over 0.25 million light years makes it very much an intergalactic wanderer being twice as far away as our own galaxy's satellites, the Magellanic Clouds. One of the most comprehensive studies of C25 was carried out by Harris et al. using the Hubble Space Telescope (1997). They concluded that C25 is perhaps the most important cluster in the outer Milky Way halo. It is much more luminous than other outer halo clusters and is in fact amongst the 5 most luminous in our galaxy. Its large core radius and half mass radius are unlike any inner cluster and demonstrate that it cannot have formed in the inner halo and has somehow migrated out on a highly elliptical orbit. It has a very low metallicity and its age is a crucial factor in understanding the formation of the Milky Way itself. C25 was formed about 14 to15 billion years ago in an era when our galaxy was beginning to make its first stars, in a region around 700,000 light years across.

Galaxy in Canes Venatici

Pedro Re

Name and/or Catalogue Designations:
NGC 4244
H V 41
UGC 7322
IRAS12150+3804
PGC 39422

Type of object: Galaxy

Catalogue position for epoch J2000.0
Right ascension: 12h 17m 30.0s
Declination: +37° 48′ 27″

Constellation: Canes Venatici

Object information:
Magnitude: 10.4
Size: 18.5′ × 2.3′
Position angle: 48°
Object classification: Scd – SA(s)cd
NGC Description: pretty bright, very large, extremely extended P.A. 43°, very suddenly
    bright to the middle, very small bright nucleus or star?
Note: classic edge-on Scd

Star atlas chart numbers:
Millennium Star Atlas, Charts 633–634, Volume II
Sky Atlas 2000.0, Chart 7
Uranometria 2000, Chart 107, Volume 1

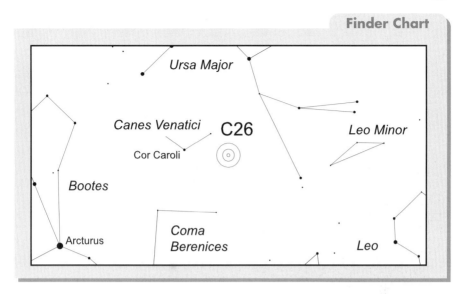

Visual Description

There are several slender edge-on galaxies in the Caldwell Catalogue. This is only to be expected as they are visually very rewarding, but C26 must be the thinnest. It is incredibly needle-like with a length of nearly one third of a degree yet with a width of less than 2 arc-minutes. An 8 inch (200 mm) telescope is sufficient to see this – larger telescopes surprisingly not adding much to the view – just a few foreground stars. The dust lane is really too subtle for even a 16 inch telescope to see – just a little mottling near the nucleus being evident. In all telescopes the nucleus is small and star-like with the galaxy gradually fading away from it. It is aligned north-east to south west. It is probably easiest to find from Alpha Canum Venaticorum (Cor Caroli – see C29) which, at magnitude 2.9, is the brightest in the vicinity. C26 is 7.5° west and 0.5° south of the star.

Object Description

Although almost exactly edge-on, C26 is noticeably different to the classic edge-on galaxies such as C23 and C38. The reason for the difference is that unlike those two galaxies C26 is a type Sc (or even Sd by some). Of this classification M33, the Pinwheel Galaxy in Triangulum, is the best known and is fairly typical. They are relatively low mass with "loose" arms and a tiny nucleus. As a result when seen edge-on, as is the case here, a thick nuclear bulge and prominent uniform dust lane are absent. It does not have an active nucleus and X-ray images (Harris, 1990) show no central condensation. Several research papers have been published on this galaxy, in particular by Rob Olling. In one, using observations (1996) with the Very Large Array Radio Telescope (VLA) of the 21-cm spectral line of neutral atomic hydrogen, he found that NGC 4244's dark matter halo is probably highly flattened. In another paper (also 1996) he determined C26's rotation curve, finding it to be one of the few galaxies with a falling rotation curve, Keplerian fashion. C26 is relatively nearby at a distance of 10 to 12 million light years.

Bright Nebula in Cygnus

Image

Tim Puckett

Database

Name and/or Catalogue Designations:
Crescent Nebula
NGC 6888
H IV 72

Type of object: Bright Nebula

Catalogue position for epoch J2000.0
Right ascension: 20h 12m 12.0s
Declination: +38° 20′ 00″

Constellation: Cygnus

Object information:
Magnitude: 7.5
Size: 20.0′ x 10.0′
Object classification: emission
NGC Description: faint, very large, very much extended, double star attached
Notes: central Wolf-Rayet star WR136 (HD192163)

Star atlas chart numbers:
Millennium Star Atlas, Charts 1149–1150, Volume III
Sky Atlas 2000.0, Chart 9
Uranometria 2000, Chart 84, Volume 1

Finder Chart

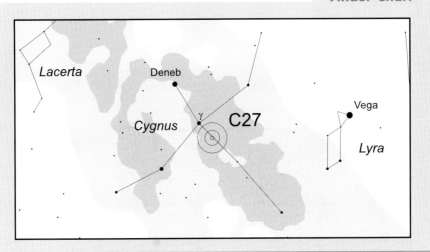

Visual Description

Under the right conditions this nebula is remarkably clear in even an 8 inch (200 mm) telescope and in a 16 inch (400 mm) several outer faint components become detectable. Those right conditions? Well the answer is an O-III filter, without it the 8 inch will struggle to see anything. With it, the crescent shape, an elliptical ring with a gap on the south-east side, and a fainter middle should all be revealed. There are reports that a UHC filter also helps. In a 16 inch telescope the nebula filters are still beneficial and essential from suburban locations. It is located 2.75° SW from Gamma Cygni, in the direction of Eta.

Object Description

Erroneously referred as a supernova remnant in some early literature, C27 is now reasonably well understood and is classed as a "Wind Blown Wolf-Rayet Ring Nebula". The explanation (Kwitter and Mac Low et al.) is that a fast rarefied wind from a Wolf-Rayet (WR) star has swept up the preceding much slower and denser wind from its supergiant stage. These interactions, at the shock-fronts, produce the ring shaped filaments we observe. It is generally believed that WR stars are evolved massive stars (masses between 5 and 50 Suns) with large mass loss rates and temperatures between 20,000 and 50,000°K. The nebula has also been studied in X-rays (Wrigge et al, 1994 using Rosat and Wrigge et al., 1996 using ASCA). They found that the interaction front, referred to above, produces large amounts hot X-ray emitting gas. The X-ray emission is outlined by the brightest optical parts of the nebula. They did however report that theoretical models did not totally agree with their findings. In 1997, Moore et al., using Hubble observations, found wind velocities of up to 2,000 kilometres per second emanating from the central star and they were able to resolve the ionisation structure of individual filaments in the fragmented shell. The distance to C27 is around 5,000 light years (Wendker et al., 1975).

C28

Image

Buil-Thouvenot

Database

Name and/or Catalogue Designations:
NGC 752
H VII 32
Collinder 23
Melotte 12

Type of object: Open Cluster

Catalogue position for epoch J2000.0
Right ascension: 01h 57m 48.0s
Declination: +37° 41′ 00″

Constellation: Andromeda

Object information:
Magnitude: 5.7
Size: 50.0′
Number of stars: 60
Magnitude of brightest star: 9.0
Object classification: III 1 m
NGC Description: cluster, very very large, rich, stars large and small, compressed
Note: old open cluster

Star atlas chart numbers:
Millennium Star Atlas, Charts 123–124, Volume I
Sky Atlas 2000.0, Chart 4
Uranometria 2000, Chart 92, Volume 1

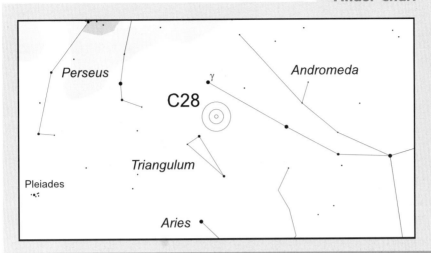

Visual Description

C28 is a beautiful splashy open cluster, even visible in binoculars as an unresolved glow. It is not a concentrated cluster but is spread out over a 1° circle, with about 60 stars on show in an 8 inch (200 mm) telescope. To see it at its best a wide-field eyepiece providing over a one degree field is essential. A 6 inch (150 mm) Richest Field Telescope at about 30× magnification is probably optimum. At the centre of the cluster is a gem in its own right, namely a multi-coloured triple star (7th,10th,10^{th}mag.). Opinions vary as to the actual colours but yellow (or orange), blue and red seem the colours to me. The cluster was probably discovered before 1654 by Hodierna (1597–1660), an astronomer at the court of the Duke of Montechiaro, and not found again until William Herschel catalogued it. It is to be found 4.8° SSW of Almach (Gamma Andromedæ).

Object Description

C28 is a very old cluster located only 1200 light years away (about 3 times the distance to the Pleiades). Mermilliod et al. (1998) have carried out an 18 year radial-velocity survey of 30 red giants in the field of cluster. The membership of 15 stars was confirmed plus 4 spectroscopic binaries. They were able to determine the orbits for three of these with periods of 3321, 127 and 5276 days. The frequency of binary stars in the cluster they found to be 27% – a fairly typical value. Various other groups have tackled determining C28's age precisely. One group (Daniel et al., 1994) used radial velocities to clarify cluster members and, by fitting standard theoretical isochrones to its colour magnitude diagram, came to an age of around 1.9 billion years. Another group, Dinescu et al. (1995) also used isochrone fitting and came to an age of 2 billion years. C28 is a source of "soft" X-rays and Belloni & Verbunt (1995) detected 49 sources of which they could identify 7 with cluster members. Three were short-period binaries, one a blue straggler and one a rapid rotator. The eclipsing binary DS Andromedae, a cluster member, was also a probable source.

Pedro Re

Database

Name and/or Catalogue Designations:
NGC 5005
H I 96
UGC 8256
IRAS13086+3719
PGC 45749

Type of object: Galaxy

Catalogue position for epoch J2000.0
Right ascension: 13h 10m 56.2s
Declination: +37° 03′ 29″

Constellation: Canes Venatici

Object information:
Magnitude: 9.8
Size: 6.3′ × 3.0″
Position angle: 65°
Object classification: SAB(rs)bc
NGC Description: very bright, very large, very much extended P.A. 66°, very suddenly
   brighter to the middle nucleus
Note: a LINER galaxy, paired with NGC 5033

Star atlas chart numbers:
Millennium Star Atlas, Charts 629–630, Volume II
Sky Atlas 2000.0, Chart 7
Uranometria 2000, Chart 109, Volume 1

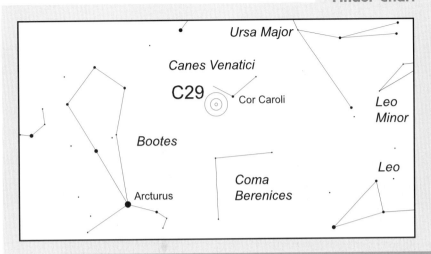

Visual Description

In a nutshell: a typical bright, magnitude 9.7, spiral galaxy about 5 × 2 arc-minutes in size with a prominent concentrated nucleus. Small telescopes, say 6–8 inches, show a definite oval shape aligned ENE to WSW, but with no detail other than the nuclear brightening. Larger telescopes show mottling but as this galaxy has no strong continuous dust lanes its spiral nature is not obvious. Within a low power field is the larger but slightly fainter (magnitude 10.6) spiral galaxy NGC5033, about 40 arc-minutes to the south-east. C29 is located 3.2° to the ESE of Alpha Canum Venaticorum (magnitude 2.9). This is of course the star, Cor Caroli, named in honour of King Charles II's return to the throne. It is an easy double with a white/blue primary and fainter reddish secondary about 20 arc-seconds away. Too good to miss on your way to C29.

Object Description

C29 is a classic spiral galaxy not unlike our own Milky Way, viewed at 23° from edge-on. The similarity also extends to morphology, both being classed as having small bars. This classification (known as SAB) was added (by de Vaucouleurs in 1959) to Hubble's original "tuning fork" diagram when it became apparent that many "typical" spirals did in fact have bars, albeit much less prominent than classical (SB) barred spirals. However, C29 is one of a type of galaxy known as LINERs, somewhat similar to Seyfert galaxies (Viegas-Aldrovandi & Gruenwald, 1990). LINER stands for Low-Ionisation Nuclear Emission-line Region and was devised by Hechman (1980) to describe a new class of emitting region in the nuclei of galaxies. The existence of a LINER in the galactic nucleus is related to the presence of a compact nuclear radio source. The source of the emission of LINERS could be from the ionising radiation caused by massive stars (called "warmers") or from the accretion disc around a black hole. In 1996 Claudio Bottari in Italy discovered a supernova (SN 1996ai) in C29, which was an extremely obscured type-Ia.

Galaxy in Pegasus

David Ratledge

Name and/or Catalogue Designations:
NGC 7331
H I 53
UGC 12113
PGC 69327

Type of object: Galaxy

Catalogue position for epoch J2000.0
Right ascension: 22h 37m 05.2s
Declination: +34° 25′ 10″

Constellation: Pegasus

Object information:
Magnitude: 9.5
Size: 11.4′ × 4.0′
Object classification: Sb
Position angle: 171°
NGC Description: bright, pretty large, pretty much extended (PA 163°), small much brighter
    to the middle
Note: three adjacent galaxies NGC 7335, 7336, 7337

Star atlas chart numbers:
Millennium Star Atlas, Charts 1141–1142, Volume III
Sky Atlas 2000.0, Chart 9
Uranometria 2000, Chart 123, Volume 1

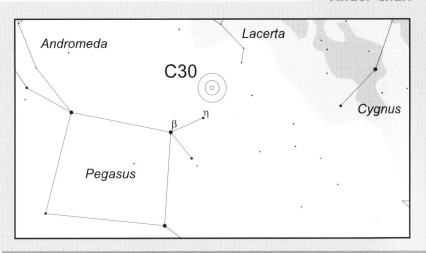

Andromeda

Lacerta

C30

Cygnus

β

η

Pegasus

Visual Description

This is the brightest of several galaxies in Pegasus and it is easy to star-hop to it using first Beta and then Eta Pegasi, then on 4.3° NNW. Being not far from edge-on it has sufficient contrast in its 10th magnitude to make it easily visible in an 8 inch (200 mm) telescope. Look for an elongated north-south smudge. Its three companions on its eastern side are, from north to south, NGC7336, 7335, 7337. They are much more difficult to observe. NGC7335 is the least difficult but at 14th magnitude really needs a 16 inch for a certain view. C30/NGC7331 is often used as the guide post to the nearby Stephan's Quintet, that enigmatic group of (allegedly) interacting galaxies with different red-shifts. This group is pretty faint however and again a 16 inch is really required just to glimpse them from all but the very darkest of skies.

Object Description

The best way to describe this spiral galaxy is the same as the Andromeda galaxy (M31) but 20 times further away. Its three companions mentioned above pose a bit of a problem. Whilst they may look like satellite galaxies their red-shifts indicate that they are probably background objects. However, the distance to C30 has been derived (Hughes et al., 1998) from Cepheid variables observed with the Hubble Space Telescope. The periods for 13 of these variable stars were established and, by comparing them with similar stars in the Large Magellanic Cloud, a distance of 50 million light years was calculated. Other recent research, by Stockdale et al. (1998) using the first deep X-ray observations of C30, has discovered a strong source coincident with the centre of the galaxy and is taken as evidence for a massive black hole there. Again in 1998, Bianchi et al. using the James Clark Maxwell Telescope detected a dust ring in C30 thought to be associated with molecular gas and star formation.

C31

Image

David Ratledge

Database

Name and/or Catalogue Designations:
Flaming Star Nebula
IC 405

Type of object: Bright Nebula

Catalogue position for epoch J2000.0
Right ascension: 05h 16m 12.0s
Declination: +34° 28′ 00″

Constellation: Auriga

Object information:
Magnitude: 6.0 (star AE Aurigae only)
Size: 37.0′ × 19.0′
Object classification: reflection & emission
NGC Description: 6.7 mag. star with pretty bright very large nebula
Notes: central star is the variable AE Aurigae

Star atlas chart numbers:
Millennium Star Atlas, Charts 113–114, Volume I
Sky Atlas 2000.0, Chart 5
Uranometria 2000, Chart 97, Volume 1

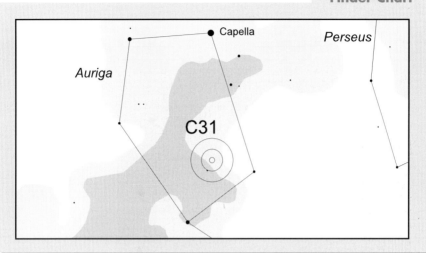

Visual Description

The Flaming Star Nebula in Auriga is a mixed refection (blue) and H-II emission (red) nebula with several components including IC405, 410 and 417 plus the variable star AE Aurigae. Whilst AE Aurigae is easy to see at magnitude 6 (see below), the nebula is not. Even in huge 20 inch (500 mm) Dobsonians it is pretty faint. Modern technology has however come to our rescue. Narrow band nebula filters, tuned to the exact emission lines of particular elements present in nebulae, do make it plainly visible. The filter that works best on C31 appears to be an H-beta, which reveals the full extent (some compare the shape to a letter q), and, to a lesser extent, UHC and O-III types. The most prominent parts of the nebula are north of AE Auriga and are, to me, visually reminiscent of the Veil Nebula (C33/34). It is located 3° WSW of the open cluster M38 and 11.5° south of Capella.

Object Description

The nebula is visible only because of a chance event. The star AE Aurigae happens to be passing through this nebula of interstellar dust and gas, and the energy emitted from the star is illuminating the nebula. Once the star passes by, the nebula will cease to shine. What caused AE Auriga to be shot through the cloud appears to be an ancient violent event (multiple supernovae?) in the nearby Orion complex. AE Auriga was ejected at a speed of 130 km/s. In this event, 53 Arietis and Mu Columbae were also flung out at slightly lesser speeds but in markedly different directions. It is thought that the shockwave from this event also caused Barnard's Loop, a giant curving nebula on the eastern side of Orion. AE Aurigae is a variable star of spectral type O9.5 with erratic variations between 5.4 and 6.1 magnitude. One controversial "discovery" was the report by Feibelman (1989) of his detection in IC405 of 60 "socket stars". These stars were apparently embedded in the nebula and surrounded by symmetrical regions of lower brightness. However, Schaefer (1994) took new CCD images and could find no evidence for sockets surrounding the stars.

 Image

Tim Puckett

Database

Name and/or Catalogue Designations:
NGC 4631
H V 42
UGC 7865
IRAS12396+3249
PGC 42637

Type of object: Galaxy

Catalogue position for epoch J2000.0
Right ascension: 12h 42m 07.7s
Declination: +32° 32′ 28″

Constellation: Canes Venatici

Object information:
Magnitude: 9.2
Size: 17.0′ × 3.5′
Position angle: 86°
Object classification: Sc or SBd
NGC Description: very bright, very large, extremely extended P.A. 70°, brighter nucleus in the middle, star attached north
Notes: NGC 4631 with NGC 4627 = Arp 281

Star atlas chart numbers:
Millennium Star Atlas, Charts 653–654, Volume II
Sky Atlas 2000.0, Chart 7
Uranometria 2000, Chart 108, Volume 1

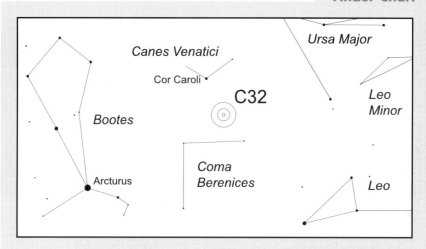

Visual Description

Visually this galaxy is very rewarding. It is a huge edge-on galaxy with a peculiar appearance, being thick at one end (east) and tapering to a point at the other (west). This has lead to it being (unofficially) named as the Whale Galaxy, a very apt description. This appearance is marked in even an 8 inch (200 mm) telescope. Larger telescopes add more detail revealing a mottled appearance with at least two prominent knots, one just east of centre and another fainter one on the other side. There is a faint 12th magnitude star on the north side, which acts as the pointer to its tiny companion galaxy NGC4627. With dark skies and averted vision, an 8 inch will just show it. In very low powers the nearby galaxy NGC4656 is also visible. C32 is located 6.5° SSW of Alpha Canes Venatici, a 3rd magnitude star.

Object Description

One only has to compare its appearance to other near edge-on galaxies C23 & C38 to realise something strange has happened to C32. It is highly disturbed due to interactions with its several companions. These comprise a large edge-on (NGC 4656) 30′ to the SE, a dwarf elliptical (NGC 4627) 3′ to the NW plus 3 fainter dwarfs. C32 is bright with star formation with a high H-alpha luminosity. Weliachew et al. (1978) and Rand (1994) detected severe tidal disruptions of its gas with 5 spectacular tidal spurs emerging from its disc. The galaxy also has an extended halo prominent in X-rays (Wang et al., 1995) and radio waves (Golla & Hummel, 1994). Perhaps the most interesting discovery in C32 was of two "supershells" in the disc of the galaxy, which was made by Rand and Van der Hulst (1993). The energy to produce these seemed to require prodigious numbers of supernova, which led Rand & Stone (1995) to suggest an alternative scenario, at least for the larger shell, which is one of the largest known. They suggested an impact, at an oblique angle, by a high velocity cloud (HVC) with the galactic disc. Certainly there is no lack of candidates given the extensive tidal debris and dwarf galaxies surrounding C32. It is about 25 million light years away.

Supernova Remnant in Cygnus

David Ratledge

Name and/or Catalogue Designations:
Veil Nebula (East)
NGC 6992-5
H V 14

Type of object: Supernova Remnant

Catalogue position for epoch J2000.0
Right ascension: 20h 56m 18.0s
Declination: +31° 42′ 00″

Constellation: Cygnus

Object information:
Magnitude: 8 approximately
Size: 60.0′ × 8.0′
Object classification: supernova remnant
NGC Description: extremely faint, extremely large, extremely extended, extremely irregular
    faint, Bifid
Note: other names are Cygnus Loop, Cirrus, Network, Bridal Veil, and Filamentary Nebula.
    Full loop size is 3.8° × 2.7°

Star atlas chart numbers:
Millennium Star Atlas, Charts 1169–1170, Volume III
Sky Atlas 2000.0, Chart 9
Uranometria 2000, Chart 120, Volume 1

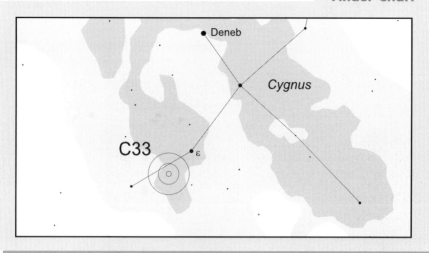

Visual Description

This, the eastern part of the Veil, is by far the easiest to see and in surprising detail. It was after all discovered by William Herschel with his narrow field of view telescope in 1784 – hardly ideal for this large object (1° long). Although detectable with binoculars in a good dark sky, an 8 inch (200 mm) telescope is realistically required for a good view. In my experience it is equally visible with and without a nebula filter. In my 12.5 inch (320 mm) telescope the curve of the Nebula is obvious and in a 16 inch (400 mm) the southern knot is evident. The nebula runs approximately north-south. The hardest part is finding this portion of the Veil. It forms an equilateral triangle with Epsilon and Zeta Cygni, approximately on the line from Gamma Cygni through Epsilon. A computer controlled telescope does make life easier!

Object Description (C33 & C34)

The Veil Nebula comprises three main parts, the Eastern Veil (C33/NGC 6992-5), the Western Veil (C34/NGC 6960) and the Northern Veil (NGC6974/6979), although there are many other portions. It is the expanding bubble of gas following a supernova explosion 15,000 to 30,000 years ago. The nebula portions we see glow where the shockwave from the blast ploughs into interstellar material. These collisions produce emissions in the radio and X-ray parts of the spectrum, as well as the optical part we see. The Hubble Space Telescope has taken detailed images (before and after its optics were corrected) of part of the Veil. Although examining only a fragment of the nebula, these images reveal some of its constituents. Detected were doubly ionised oxygen atoms (blue in the HST image) produced by the heat behind the shockwave; singly ionised sulphur (coloured red) produced well behind the shock front; and a thin hydrogen zone immediately behind the shock front (coloured green).

(continued in C34)

C34

Supernova Remnant in Cygnus

Buil-Thouvenot

Name and/or Catalogue Designations:
Veil Nebula (West)
NGC 6960
H V 15

Type of object: Supernova Remnant

Catalogue position for epoch J2000.0
Right ascension: 20h 45m 36.0s
Declination: +30° 43' 00"

Constellation: Cygnus

Object information:
Magnitude: 4.22 (star 53 Cygni)
Size: 70.0' × 6.0'
Object classification: supernova remnant
NGC Description: pretty bright, considerably large, extremely irregular faint, Kappa Cygni involved
Notes: part of the Cygnus Loop with the star 52 Cygni on the line of the nebula

Star atlas chart numbers:
Millennium Star Atlas, Charts 1169–1170, Volume III
Sky Atlas 2000.0, Chart 9
Uranometria 2000, Chart 120, Volume 1

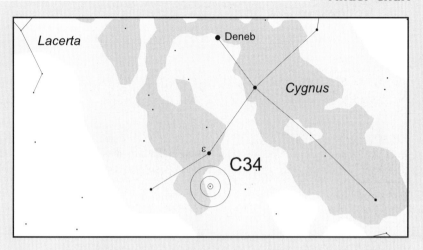

First the good news, C34 the Western Veil, is far easier to locate than C33, thanks to the presence of the star 52 Cygni (4th magnitude and a touch yellowish). Now the bad – it is definitely harder to detect, again because of that star. It is however easily in reach with an 8 inch (200 mm) telescope under reasonable conditions. Again it is extremely large and orientated roughly north-south. The portion north of 52 Cygni (below in an inverting telescope) is the brightest. From my suburban skies it required a nebula filter on my 12.5 inch (320 mm) telescope to make it clearly visible, with its crooked shape noticeable. The filter helps more by dimming the star than brightening the nebula. If you are still struggling to see it try increased magnification – doubling your lowest power should help. To locate it all that that is needed is to head 3.25° south from Epsilon Cygni (mag. 2.5) to 52 Cygni and pop in that nebula filter.

(continued from C33)
Probably the most comprehensive survey of C33/34 was undertaken by Levenson et al. (1998), who produced a complete atlas of the whole Cygnus Loop complex. Their observations were in the O-III, H-alpha and S-II wavelengths. They identified, in addition to the visible traces, non-radiative shocks around half the perimeter of the loop. They concluded that the blast wave is not breaking out of a dense cloud but is running into confining walls. The interstellar medium dominates not only the visible appearance of the nebula but also the continued evolution of the blast wave. Levenson et al. (1997) have also carried out an X-ray study of C33/34 and concluded that the supernova explosion took place in a pre-existing cavity in the interstellar medium.

The Veil lies about 2,000 light years away (estimates vary widely) and is now around 100 light years in diameter. The supernova remnant has never been found, perhaps because it totally destroyed itself in the catastrophic explosion that gave us the magnificent Veil Nebula. The alternative is that is was ejected in the explosion.

Galaxy in Coma Berenices

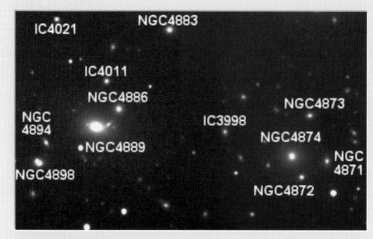

Pedro Re

Name and/or Catalogue Designations:
NGC 4889
H II 321
UGC 8110
PGC 44715

Type of object: Galaxy

Catalogue position for epoch J2000.0
Right ascension: 13h 00m 08.3s
Declination: +27° 58′ 39″

Constellation: Coma Berenices

Object information:
Magnitude: 11.5
Size: 2.8′ × 2.0′
Position angle: 80°
Object classification: E4
NGC Description: pretty bright, pretty much extended, bright to the middle, 7th magnitude
    star to north
Note: brightest in Coma Cluster (Abell 1656), cluster approximately 2° diameter with
    800 galaxies to mag. 16.5 (Zwicky)

Star atlas chart numbers:
Millennium Star Atlas, Charts 653–654, Volume II
Sky Atlas 2000.0, Chart 7
Uranometria 2000, Chart 109, Volume 1

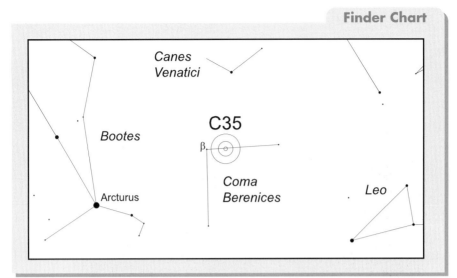

Visual Description

C35 is the brightest galaxy in the Coma Galaxy Cluster and, with its nearby companion NGC4874, is one of the two dominant members. The Coma Cluster is not to be confused with the much nearer Virgo cluster whose members are spread across Virgo and Coma Berenices. The Coma Cluster is about 7 times further away and just seeing it is a significant achievement. Under a good dark sky, C35 (and NGC4874) are visible in a 6 inch (150 mm) telescope but realistically an 8 inch is the minimum for a guaranteed sighting. C35 has a bright core, which makes it stand out more than any other cluster members. To see further members head west to NGC 4874 (mag. 11.7) or east and a little south to NGC4921 (mag. 12.2). Another galaxy, NGC4911 (mag. 12.8), lies a little south and to the west of 4921. These are fainter than their magnitudes indicate but are quite easy in a 16 inch. The cluster is 2.5° west of Beta Comæ Berenices (mag. 4.2). A 7th magnitude star is just north of NGC 4874.

Object Description

The Coma Galaxy Cluster (Abell 1656) is more than 350 million light years away, near the north galactic pole. At the cluster core are two giant elliptical galaxies, NGC 4889 (C35) and NGC 4874. These two dominate a very rich cluster of around 1000 galaxies. The cluster is nearly spherical and consists mostly of elliptical (many highly-flattened) and dust free SO type galaxies. In a recent study of redshifts in the cluster (Colless & Dunn, 1996) found evidence for the cluster to be a merger of two clusters, one each centred on NGC 4889 and NGC 4784. There is even evidence for other sub-concentrations centred on other massive galaxies in the cluster (Gurzadyan & Mazure, 1998). Backing up this merger(s) hypothesis, Trentham & Mobasher (1998) found evidence in the form of giant stellar debris that could have resulted from damaging galaxy collisions. Also in 1998, Conselice & Gallagher found what they described as "galaxy aggregates", galaxies surrounded by numerous small companions. They suggested that such objects are likely to have formed only in clusters, which are accreting galaxies.

Galaxy in Coma Berenices

Pedro Re

Name and/or Catalogue Designations:
NGC 4559
H I 92
UGC 7766
IRAS12334+2814
PGC 42002

Type of object: Galaxy

Catalogue position for epoch J2000.0
Right ascension: 12h 35m 57.8s
Declination: +27° 57′ 36″

Constellation: Coma Berenices

Object information:
Magnitude: 9.8
Size: 13.0′ × 5.2′
Position angle: 150°
Object classification: Sc
NGC Description: very bright, very large, much extended P.A. 150°, gradually brighter to the middle, 3 stars following
Notes: two faint 15.5 mag. companions to SE (NGC4559 A & B)

Star atlas chart numbers:
Millennium Star Atlas, Charts 653–654, Volume II
Sky Atlas 2000.0, Chart 7
Uranometria 2000, Chart 108, Volume 1

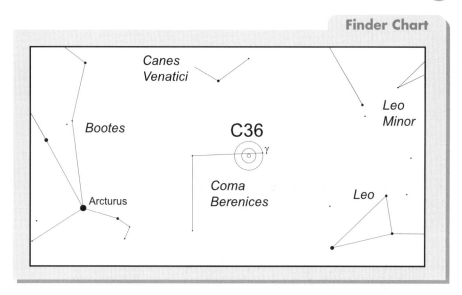

C36 is a neglected member of the Virgo cluster of galaxies (although it is Coma Berenices!) but it is nonetheless one of the most rewarding. Even a 6 inch (150 mm) telescope shows an elongated hazy patch with a slightly brighter middle. There is an 11.5 magnitude star overlaying the galaxy on the eastern side, which is also just visible in this size of telescope. An 8 inch adds a couple more stars but starts to hint at the mottled nature of this galaxy. This mottling becomes pronounced in a 16 inch and is no doubt due to the dust lanes, H-II regions and star fields present in this galaxy. It is relatively easy to find. Either locate the edge galaxy C38 (NGC4565) and then head 2° north or star-hop from Gamma Comae Berenices (a wide 4th and 7th mag. double) about 2° east.

Viewed about half-way between edge-on and face-on, this Sc galaxy has been shown, on a beautiful colour image taken with the 1.5 meter telescope on Mount Palomar, to contain prominent areas of older yellow stars, red H-II regions, and clusters of hot blue stars. A 1997 paper by Vogler et al. reported on X-ray emission from C36 observed with the ROSAT orbiting X-ray observatory. Seventeen point sources were detected of which seven were within the main ellipse of the galaxy. The most luminous source was associated with the nucleus of the galaxy (although fractionally offset from the optical nucleus) and was slightly extended, suggesting a superposition of several compact sources or diffuse emission surrounding a compact central source. Another bright source was found in an outer spiral arm and as there is little radio or infra-red emission from this region they suggested that the most likely source was a 100 year old supernova remnant. In 1941 there was a type II supernova in this galaxy of magnitude 13.2 (SN 1941A) but this was a different source. No emission was detected from a possible halo.

Open Cluster in Vulpecula

Image

David Ratledge

Database

Name and/or Catalogue Designations:
NGC 6885
H VIII 20
Collinder 417

Type of object: Open Cluster

Catalogue position for epoch J2000.0
Right ascension: 20h 12m 00.0s
Declination: +26° 29′ 00″

Constellation: Vulpecula

Object information:
Magnitude: 8.1
Size: 7.0′
Number of stars: 30
Magnitude of brightest star: 6.0 (20 Vulpeculae)
Object classification: III 2 p
NGC Description: cluster, very bright, very large, rich, little compressed, stars 6th to 11th
    magnitudes
Note: includes the star 20 Vulpeculae and forms a double cluster with NGC 6882,

Star atlas chart numbers:
Millennium Star Atlas, Charts 1193–1194, Volume III
Sky Atlas 2000.0, Chart 9
Uranometria 2000, Chart 163, Volume 1

Finder Chart

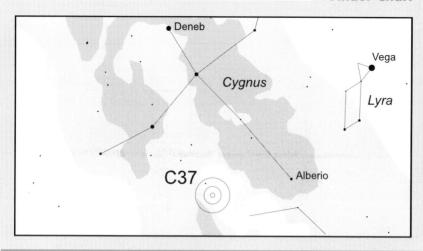

Visual Description

I suspect that not many observers, other than Patrick Moore, knew about this open cluster. It is an easy binocular object (10 × 50s) but lying to the east of the spectacular Cygnus star cloud many have passed it by. It is a shame because it is worth seeking out. In binoculars it is a small hazy patch, about 1/5th of a degree in size, mainly to the south-west of the star 20 Vulpeculae. In a 6 inch (150 mm) telescope this haze is revealed as several 9th magnitude stars with the cluster fairly spread out and centred on 20 Vul. The stars to the north-west of this 6th magnitude star are part of the bigger cluster NGC6882, which C37 overlies. Taken with a background of Milky Way stars unravelling the two clusters is very tricky! The glare of 20 Vul also makes detection of the smaller C37 more difficult. To locate this cluster either head 9° ENE from Albireo (Beta Cas) or 4.75° due north of the Dumbbell Nebula (M57). In both cases look for a little triangle of 6th magnitude stars (18, 19 & 20 Vul)

Object Description

Three objects for the price of one! Two open clusters, C37 (NGC6885) and NGC6882, plus the variable star 20 Vulpeculae. C37 is thought to be a very old cluster and Phelps et al. (1993) in their search for the Milky Way's oldest clusters thought it warranted further study but had to reject it because of insufficient data. Its age would appear to be about 1 billion years. Its distance also is not well established but is around 1900 light years. However, the European Space Agency's Hipparcos mission has established a reliable distance to 20 Vulpeculae. Its distance turns out to be 1140 ± 211 light years. Even taking the upper range this places 20 Vulpeculae in front of C37 and therefore probably not a true cluster member. But in view of the uncertainty in C37's distance even this is not definite. The photometry measurements by Hipparcos (the Tycho catalogue) gave 20 Vulpeculae a spectral class of B7Ve and with magnitude variations of 5.87 to 5.92.

C38

Image

David Ratledge

Database

Name and/or Catalogue Designations:
Needle Galaxy
NGC 4565
H V 24
UGC 7772
PGC 42038

Type of object: Galaxy

Catalogue position for epoch J2000.0
Right ascension: 12h 36m 20.6s
Declination: +25° 59′ 05″

Constellation: Coma Berenices

Object information:
Magnitude: 9.6
Size: 15.5′ × 1.9′
Position angle: 136°
Object classification: Sb
NGC Description: very remarkable, bright, extremely large, extremely extended P.A. 135°,
    very small brighter nucleus in the middle equal to a star mag. 10 to 11
Note: inclination 4° from edge-on, galaxy NGC 4562 is 15 arc-minutes to SE

Star atlas chart numbers:
Millennium Star Atlas, Charts 677–678, Volume II
Sky Atlas 2000.0, Chart 7
Uranometria 2000, Chart 149, Volume 1

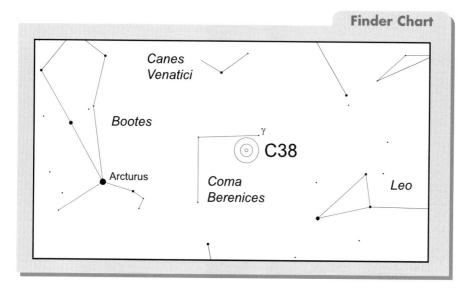

Visual Description

Circumstances have rendered C38 remarkably easy to see. Its orientation almost exactly edge-on provides plenty of contrast and its location far away from the plane of the Milky Way means there is little obscuration and no distracting star-fields to get in the way. Even a 6 inch (150 mm) shows a slender ghostly spindle plus the stellar-like centre. However, to detect that famous dust lane (easiest on the NW edge near the nucleus) at least an 8 inch (200 mm), and ideally a 12 inch (300 mm), telescope is required. The nearby galaxy, NGC4562, is 14th magnitude and requires a 16 inch telescope to spot it. C38 is located to the east of Coma Berenices "hair", the loose cluster Melotte 111. A 50 mm finder should spot it about 3° SE of Gamma Coma Berenices.

Object Description

The largest of the edge-on spirals, C38 is located approximately 30 million light years distant. It is one of the more massive nearby galaxies (Huchtmeier et al., 1980). Thought to be very reminiscent of our own Milky Way with the dark dust lane through the middle being equivalent to our galaxy's Great Rift, which divides the Milky Way in two. When the centre of C38 was observed with the Hubble WFPC2, the dark lane, which is so prominent in ground-based images, was resolved into numerous clouds. These mark the locations of relatively dense concentrations of interstellar gas and dust. However, unlike dark clouds in the Milky Way or other nearby galaxies, those in NGC 4565 seem to be producing few bright, massive stars. Another interesting feature of C38 is that it shows bending or warping at its edges. Other galaxies with warped discs are known but are quite rare and warps are generally thought to be caused by a tidal interaction with a nearby companion. However, in C38's case, there is no companion near enough. Possibly the most unusual observations of C38 were made using a rocket-borne infra-red telescope (Uemizu et al., 1998). They found that, unlike similar edge-on galaxy NGC5907, C38 did not appear to have a visible halo.

C39

Planetary Nebula in Gemini

2x enlargement

N

David Ratledge

Name and/or Catalogue Designations:
Eskimo Nebula or Clown Face Nebula
NGC 2392
H IV 45
PK 197+17.1

Type of object: Planetary Nebula

Catalogue position for epoch J2000.0
Right ascension: 07h 29m 09.4s
Declination: +20° 54' 45"

Constellation: Gemini

Object information:
Magnitude: 8.6
Size: 47.0" × 43.0"
Magnitude of central star: 9.8
Object classification: 3b
NGC Description: bright, small, round, 9th magnitude star, 8th magnitude star north
    following 100"
Note: spectral type of central star O7f

Star atlas chart numbers:
Millennium Star Atlas, Charts 175–176, Volume I
Sky Atlas 2000.0, Chart 5
Uranometria 2000, Chart 139, Volume 1

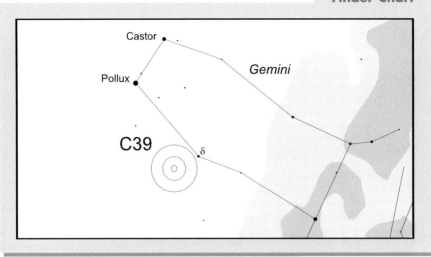

The Eskimo Nebula is one of the northern hemisphere's winter favourites and probably first got this name from its appearance in photographs taken in red light with the Palomar 200 inch (5 metre) telescope. Although very small (about the size of Jupiter's disc) it is bright and unusually is one object where high magnification improves the view. Below 100× it is just a bluish dot, a bit like a tiny globular cluster. Doubling magnification in just an 8 inch (200 mm) telescope it starts to get interesting! The central star appears and a dark arc is faintly discernible on the "face" towards the north-west. In a 16 inch telescope, again at high power (300×), the markings on the face become more pronounced. The outer halo ("fur hood") is very faint and requires averted vision in all but the largest telescope. It is located 2.3° ESE of Delta Geminorum (Wasat).

C39 was discovered by William Herschel in 1787, who thought the nebula surrounding the central star was its "atmosphere". Lord Rosse (and his assistant Storey) also observed it with the 6 foot telescope and could not "resolve it". We now know C39 is a round planetary Nebula (PN) with a prominent inner ring (the face) and a more diffuse outer halo (the hood). However the circular morphology conceals much underlying complexity with bars, wisps and knots. Most theoretical studies have concluded that C39 is an elliptical PN seen almost pole-on. Although their formation is now reasonably well understood (see C2) there is one aspect where even today we struggle. That is their distances. Their determination remains a tricky problem, despite some recent attempts by several researchers employing a wide variety of methods. Estimates of the distance to the Eskimo have ranged from 1,600 light years (Zhang, 1993) to 6,500 light years (several teams). A recent paper by Marciel and Cazetta (in press), using a "gravity" method, has produced evidence that the distance to C39 is possibly even higher at 7,500 light years. The team of Hajian et al. (1995) used expansion algorithms to derive a distance of 4,500 light years, which is probably a good compromise value.

Galaxy in Leo

Buil-Thouvenot

Name and/or Catalogue Designations:
NGC 3626
H II 52
UGC 6343
PGC 34684

Type of object: Galaxy

Catalogue position for epoch J2000.0
Right ascension: 11h 20m 03.7s
Declination: +18° 21′ 30″

Constellation: Leo

Object information:
Magnitude: 11.0
Size: 2.8′ × 2.0′
Position angle: 157°
Object classification: Sb
NGC Description: bright, small, very little extended, small bright middle
Note: galaxy UGC6341 is 5 arc-minutes to south and the pair NGC3607/3608 is
    45 arc-minutes to west

Star atlas chart numbers:
Millennium Star Atlas, Charts 703–704, Volume II
Sky Atlas 2000.0, Chart 13
Uranometria 2000, Chart 146, Volume 1

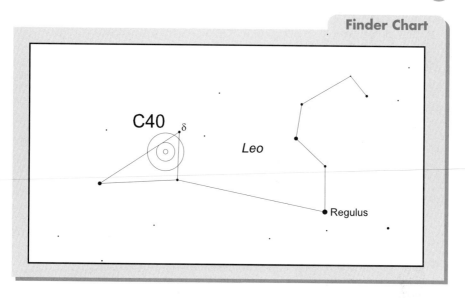

Because of C40's proximity to several brighter galaxies, including two Messier ones (M65/66), this galaxy has been much overlooked and is virtually unknown. I must admit that I had to search it out especially never having seen it before. In addition, when looking for it, I first mis-identified it finding the slightly brighter NGC3607 initially, which is in fact 48 arc-minutes away to the WSW. This galaxy is fully one magnitude brighter (mag. 10) and without it I doubt whether I would have been able to find C40, which is very insignificant visually. It just appears as a faint small circular patch even in my 16 inch (400 mm) telescope. It is located 2.75° SE of Delta Leonis. To star-hop to it I would recommend first locking on to M65 and then heading 5° almost due north to NGC 3607. C40 is then 48 arc-minutes ENE. Note the 11th magnitude galaxy about 6 arc-minutes north of NGC3607 is not C40 but NGC3608.

C40's interest lies not in its visual appearance but in the fact that it is one of the few galaxies with counter-rotating molecular gas, dubbed "multispin" galaxies. In fact the team of Garcia-Burillo et al. (1998), who made the discovery, claimed that it was the first time such a discovery had been made in a *spiral* galaxy. Using a 30 metre radio telescope they detected huge amounts of molecular and ionised gas rotating around the galaxy in the opposite direction to that of the stars. Surprisingly no evidence (including optical and infra-red observations) could be found for the resulting star formation. Where all the wrong-way material came from could not be pinned down for certain but the likely scenario is that C40 recently swallowed up a giant gas cloud – one equivalent to a billion suns. The gas disc rotation was just too uniform to have resulted from any violent large scale shocks. Whichever way it must have been a very recent event as in such an occurrence the gas cloud should be rapidly turned into new stars.

C41

Open Cluster in Taurus

Image

David Ratledge

Database

Name and/or Catalogue Designations:
Hyades
Collinder 50
Melotte 25

Type of object: Open Cluster

Catalogue position for epoch J2000.0
Right ascension: 04h 27m 00.0s
Declination: +16° 00′ 00″

Constellation: Taurus

Object information:
Magnitude: 0.5
Size: 5°
Number of stars: 40
Magnitude of brightest star: 3.4
Object classification: II 3 m
NGC Description: very very bright, very very large, irregular
Note: Taurus moving cluster, Aldebaran not a member

Star atlas chart numbers:
Millennium Star Atlas, Charts 185–186, Volume I
Sky Atlas 2000.0, Chart 11
Uranometria 2000, Chart 178, Volume

Finder Chart

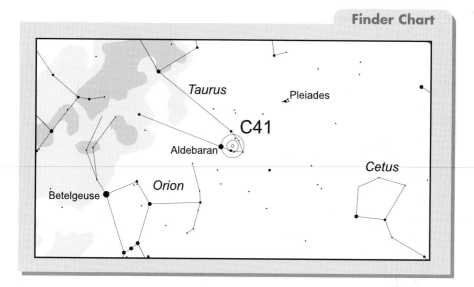

Visual Description

If the Messier list included the Pleiades then it is only right that the Hyades are not left out and are included in the Caldwell collection! It is a true cluster, with the exception that Aldebaran is a foreground star and only "in" the cluster due to a chance alignment. C41 is one of the few objects visible from city skies and binoculars provide by far the best view. Spanning over 5° and with over 100 stars brighter than 9th magnitude, 10 × 50 binoculars are perfect. In a telescope the cluster all but disappears.

Object Description

As the nearest moderately rich star cluster, the Hyades has played an important role in astrophysics for more than a century. Star clusters are crucial for understanding the lives of the stars everywhere, because all the members of a cluster formed at the same time from the same raw materials. The distance of the Hyades is also the starting point for astronomical distance measurements, which extend throughout the galaxy and beyond. The accurate determination of which is essential to determining the overall distance scale and age of the Universe. But the accuracy of the theories has been limited by uncertainties in the Hyades' distance. However, thanks to measurements by the European Space Agency's star-mapping satellite Hipparcos, the uncertainty is over (Brown & Perryman, 1997). Its distance turns out to be 151 light years to an accuracy of 1% and its age 625 million years. By combining parallaxes and proper motions from Hipparcos, together with ground based radial (line of sight) velocities, the position and velocity components of cluster members were calculated establishing the 3-dimensional structure of the Hyades. This showed that the individual stars of the Hyades are bound together by the gravity of the cluster as a whole with relatively massive stars having sunk towards the cluster's centre of gravity, but with some of the lighter ones escaping the cluster. They also found 13 more likely cluster members.

Globular Cluster in Delphinus

David Ratledge

Name and/or Catalogue Designations:
NGC 7006
H I 52

Type of object: Globular Cluster

Catalogue position for epoch J2000.0
Right ascension: 21h 01m 29.5s
Declination: +16° 11' 15"

Constellation: Delphinus

Object information:
Magnitude: 10.6
Size: 2.8'
Object classification: 1
NGC Description: bright, pretty large, round, gradually brighter middle
Notes: very remote globular.
Distance from the Sun: 40.7 kpc.
Distance from the galactic centre: 38.0 kpc.

Star atlas chart numbers:
Millennium Star Atlas, Charts 1215–1216, Volume III
Sky Atlas 2000.0, Chart 17
Uranometria 2000, Chart 164, Volume 1

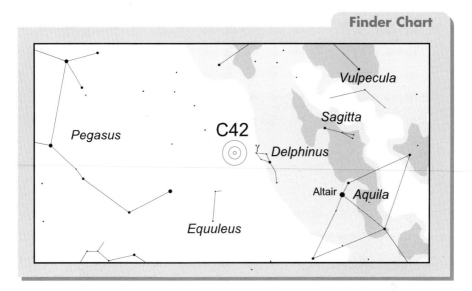

Visual Description

This enigmatic globular cluster is one of the Milky Way's most remote, although C25 and several of the obscure Palomar clusters are further away. It is surprisingly bright at magnitude 10.6 (total) and is quite an easy object in only a 6 inch (150 mm) telescope. Large 20 × 80 binoculars just about reach it with averted vision. In fact bigger telescopes do not reveal much more until they are of sufficient aperture to start resolving the individual stars. My 16 inch was unable to resolve them although there was some evidence of "clumpiness". I suspect that one of those giant Dobsonians (20 inch and upwards) would be required for definite resolution, as has been reported (Lew Gamer – 20 inch Dobsonian). In more normal sized telescopes a small (about 2 arc-minutes) circular haze with a definite brightening to the centre is discernible. Easy to find as it is located fifteen arc-minutes due east of Gamma Delphini.

Object Description

C42 is a remote globular cluster far out into the galactic halo (130,000 light years distant from us) and its importance was first established by Allan Sandage and Wildey in 1967. They found that its stars were moderately metal poor (elements higher than helium) but unusually its "horizontal branch" consisted of almost entirely of red stars. This is generally only the case in those clusters with a considerably higher metal abundance. This is the infamous "second parameter" problem. What is it that varies from cluster to cluster, other than metallicity, that affects the morphology of the horizontal branch? Cohen and Frogel (1981) again examined C42 to see if they could resolve this puzzle. They found that the second parameter was not Carbon, nitrogen and oxygen abundancies. More recently Kraft et al. (1998) used the power of the giant Keck telescope to examine red giants in C42. They confirmed that it had similar elemental ratios to other halo globular clusters but its horizontal branch did indeed have a large proportion of red stars. Their conclusion was that deep mixing had taken place inside the cluster's red giant stars and this could perhaps be the "second parameter".

Image

Pedro Re

Database

Name and/or Catalogue Designations:
NGC 7814
H II 240
UGC 8
PGC 218

Type of object: Galaxy

Catalogue position for epoch J2000.0
Right ascension: 00h 03m 15.1s
Declination: +16° 08′ 45″

Constellation: Pegasus

Object information:
Magnitude: 10.6
Size: 6.0′ × 3.0′
Position angle: 135°
Object classification: Sb or S0
NGC Description: considerably bright, considerably long, very gradually brighter to the
    middle
Notes: nearly edge-on with equatorial dust lane

Star atlas chart numbers:
Millennium Star Atlas, Charts 197–198, Volume I
Sky Atlas 2000.0, Chart 10
Uranometria 2000, Chart 125, Volume 1

Finder Chart

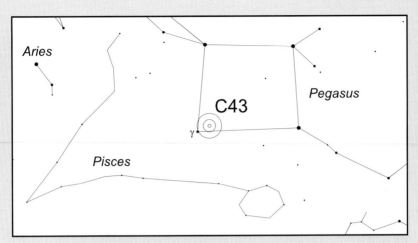

Aries

Pegasus

C43

γ

Pisces

Visual Description

Located inside the Square of Pegasus, this galaxy has many visual similarities to the more familiar Sombrero Hat Galaxy (M104). Most observers agree that it is fairly bright and easy in an 8 inch (200 mm) telescope. Where observers disagree is in seeing the dust lane, which is so prominent in images. In my experience an 8 inch is not sufficient and only shows the brighter circular middle directly, with the elongated outer areas needing averted vision. The dust lane is very fine and is just discernible in my 16 inch. Higher powers (200×) seemed to help with this. To find C43, first locate Gamma Pegasi (Algenib), which is the south-east corner of the "Square of Pegasus". Then head 2.5° WNW – look out for a 7th magnitude star, which is 12 arc-minutes WNW of C43. Now, can you see the dust lane?

Object Description

Almost exactly edge-on with a pronounced central bulge and a uniform slender dust lane, C43 is not unlike M104. In images and photographs the dust lane is often burnt out (over exposed) near the nucleus. The bulge was investigated by Lequeux et al. (1995) using the 3.6 metre telescope at the European Southern Observatory. They found that the outer bulge becomes surprisingly redder with increasing radius. This, they postulated, could indicate that the outer regions contain only very low mass stars. They also observed a distinct skewed warp in the stellar disc of C43. Warps have been noticed in a few spiral galaxies, notably M31. Van der Kruit & Searle (1982) observed this warp in H-I line observations of the outer disc. This group also determined C43's distance at around 50 million light years. The globular clusters of C43 have also been studied by Bothun et al. (1992). Down to a magnitude of 24.5 they detected around 120 globular clusters. Correcting for incompleteness they estimate that the galaxy probably has a total population of around 500, which is a high value more typical of elliptical galaxies.

C44

Image

Pedro Re

Database

Name and/or Catalogue Designations:
NGC 7479
H I 55
UGC 12343
IRAS 23024+1203
PGC 70419

Type of object: Galaxy

Catalogue position for epoch J2000.0
Right ascension: 23h 04m 57.1s
Declination: +12° 19′ 18″

Constellation: Pegasus

Object information:
Magnitude: 10.9
Size: 4.4′ × 3.4′
Position angle: 25°
Object classification: SBbc
NGC Description: pretty bright, considerably large, much extended P.A. 12°, between
    2 stars
Notes: excellent barred spiral

Star atlas chart numbers:
Millennium Star Atlas, Charts 1233–1234, Volume III
Sky Atlas 2000.0, Chart 17
Uranometria 2000, Chart 213, Volume 1

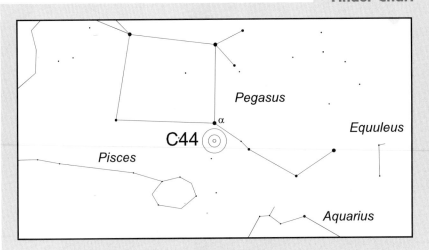

Visual Description

C44 is probably the best barred spiral galaxy in the northern hemisphere and, at a visual magnitude of 10.8, is bright enough to be visible in modest telescopes. Orientated almost face-on, an 8 inch (200 mm) telescope shows a cigar shaped object about 4 arc-minutes long by 1 wide, running almost due north-south. This is C44's prominent bar and averted vision shows some mottling in it, possibly due to the dust lanes in the bar. However the two spiral arms, which emanate from each end of the bar, require more aperture. A 16 inch reveals the western arm but even in this telescope the eastern one is still invisible. The nucleus is however obvious and looks like a 12th magnitude star. C44 is very easy to find, being 3° due south of Alpha Pegasi (Markab).

Object Description

C44 is the classical barred galaxy, almost face-on with the spiral arms winding out from a very prominent bar. As a result C44 has been extensively studied in order to understand bar formation. Also in its favour is its isolation from other galaxies and its morphology is therefore due solely to itself rather than any galactic interactions. So understand C44 and take a large step to understanding barred spirals. To do this it is necessary to study its interstellar material (and C44 has plenty) as well as just its stars. Laine et al. (1998 & 1999) used radio telescopes primarily to detect emissions from CO (carbon monoxide molecules), which are easier to observe than H-alpha but are believed to be accurate tracers of it. CO emissions were detected only in the dust lane that runs the length of the bar. Results indicate that the motion of the molecular gas is circular in the circum-nuclear region but outside this there is a strong radial velocity, inward towards the nucleus. Laine & Gottesman (1997) proposed that the more prominent western arm was the result of a minor merger. Rozas et al. (1999) on the other hand examined C44 in H-alpha light detecting over 1,000 H-II regions with only a small number in the bar. The distance to C44 is around 100 million light years.

C45

Image

Pedro Re

Database

Name and/or Catalogue Designations:
NGC 5248
H I 34
UGC 8616
IRAS13350+0908
PGC 48130

Type of object: Galaxy

Catalogue position for epoch J2000.0
Right ascension: 13h 37m 31.9s
Declination: +8° 53′ 08″

Constellation: Bootes

Object information:
Magnitude: 10.3
Size: 6.8′ × 5.0′
Position angle: 110°
Object classification: Sc or SAB(rs)bc I-II
NGC Description: bright, large, extended P.A. 150°, pretty small bright middle, round nucleus
Notes: 15th magnitude companions UGC8575 & 8629

Star atlas chart numbers:
Millennium Star Atlas, Charts 745–746, Volume II
Sky Atlas 2000.0, Chart 14
Uranometria 2000, Chart 196, Volume 1

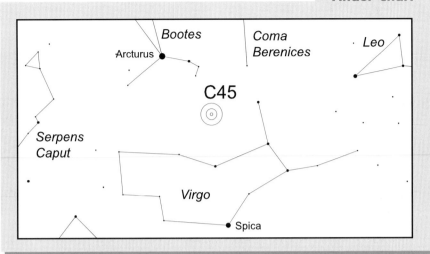

Visual Description

The constellation of Bootes is not noted for deep sky objects (it does have a few pretty double stars) but there is one object worth searching out, the spiral galaxy C45 (NGC5248). It is bright (magnitude 10.3 and relatively large (7 × 5 arc-minutes). Its central core stands out and is even visible in good finderscopes. A 6 inch (150 mm) shows it easily. The outer haze, which is in fact the spiral arms, is just visible with averted vision. To see the two main arms distinctly, increased aperture is essential. In my experience a 16 inch telescope is needed and even then they need careful scrutiny – they don't jump out like they do in images! The brightest parts of the arms are north-west and east of the galaxy's centre. The most prominent dark lane (i.e. between the arms) is north of the core. Within half a degree of C45 are the 15th magnitude galaxies UGC8575 (west) and UGC8629 (south-east). These presumed companions require dark skies and big apertures. C45 is located just inside the border with Virgo, 13.75° south-west from Arcturus.

Object Description

Originally classed as an Sc spiral galaxy (e.g. similar to M33) it has been re-classified as SAB, meaning it has a small bar. Viewed not quite face-on, it has a very bright extended core (hiding the bar) and two bright main spiral arms. Both the core and outer regions have been the subject of research. The team of Elmegreen et al. (1997) studied the central region, in particular "circum-nuclear hotspots". These are approximately evenly spaced around a circum-nuclear ring, with a spacing consistent with their formation by large-scale gravitational collapse along the periphery of the ring. The team of Chromey et al. (1996) on the other hand studied the faint spiral structure in C45. They mapped the extent of the arms and deduced that the strong outer spiral features are composed of a relatively young stellar population. There is also a published Hubble Space Telescope image of this galaxy taken in the ultra-violet.

Bright Nebula in Monoceros

Pedro Re

Database

Name and/or Catalogue Designations:
Hubble's Variable Nebula
R Mon.
NGC 2261
H IV 2

Type of object: Bright Nebula

Catalogue position for epoch J2000.0
Right ascension: 06h 39m 12.0s
Declination: +8° 44' 00"

Constellation: Monoceros

Object information:
Magnitude: 10
Size: 2.0' × 1.0'
Object classification: reflection/emission
NGC Description: bright, very much extended PA 330, nucleus cometic equal to an
    11th mag. star
Note: distance estimated at 2500 light years

Star atlas chart numbers:
Millennium Star Atlas, Charts 227–228, Volume I
Sky Atlas 2000.0, Chart 12
Uranometria 2000, Chart 182, Volume 1

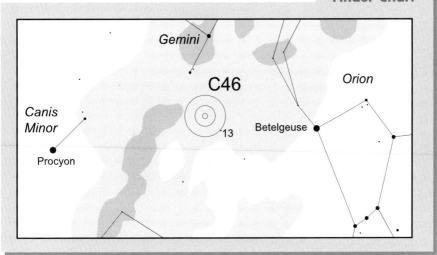

Visual Description

Discovered in 1783 by William Herschel, this object's bright cometry appearance often leads to it being mistaken for a comet. C46 is easily found in a 6 inch (150 mm) telescope and its high surface brightness means it bears magnification well. This is one of the few nebulous objects that is easily visible from suburban skies. The southern tip of the nebula is the brightest and is itself known as the variable star R Mon, although the actual star itself is invisible and buried inside the nebula (see below). To star-hop to this object, first locate the pair of stars 13 and 14 Monoceroti, and the nebula is then 1° north of 14.

Object Description

When Edwin Hubble discovered, in 1916, that this nebula was variable and, in addition, it varied in periods as short as a few weeks, it became one of the most intensely studied objects. If the nebula was really changing so quickly then speeds approaching the velocity of light would be required. When the new Hale 200 inch Telescope was completed, this was the first object to be photographed. What it is now believed to be is an intensely hot young star vigorously clearing out the material from which it formed. Gas carried north from the star has formed the comet shaped reflection nebula we can see. Gas carried south is obscured from our view by opaque material, probably an accretion disc. Those fast changes in the nebula mentioned earlier? They are thought to be merely shadow effects and the groups of Abbott et al. (1993) and Martin et al. (1994) postulated that the shadows were due to dark bodies orbiting R.Mon. Recent infra-red observations (Close et al., 1997), in conjunction with Hubble optical images have revealed that R Mon is a binary star. The primary is hot and massive (Ae/Be type) with a dusty accretion disc around 200 A.U. in diameter. The secondary is a classical T Tauri type and is at a separation of 0.69". It is very young at around 300,000 years old. The nebula itself they described as "a thin bipolar parabolic shell of dust".

C47

Globular Cluster in Delphinus

Image

David Ratledge

Database

Name and/or Catalogue Designations:
NGC 6934
H I 103

Type of object: Globular Cluster

Catalogue position for epoch J2000.0
Right ascension: 20h 34m 11.6s
Declination: +7° 24′ 15″

Constellation: Delphinus

Object information:
Magnitude: 8.9
Size: 2.0′
Object classification: 8
NGC Description: bright, large, round, well resolved, stars 16th magnitude and less
Note: 9th magnitude star 2′ to west.
Distance from the Sun: 15.2 kpc.
Distance from the galactic centre: 12.4 kpc.

Star atlas chart numbers:
Millennium Star Atlas, Charts 1265–1266, Volume III
Sky Atlas 2000.0, Chart 16
Uranometria 2000, Chart 209, Volume 1

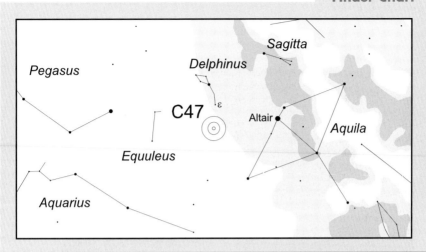

Visual Description

We have already met one globular cluster in Delphinus, the better known C42 (NGC7006) but this constellation has a brighter and larger one, namely C47 (NGC 6934). It is brighter and larger because it is so much closer (see below). In an 8 inch (200 mm) telescope it is perfectly circular and the centre is quite bright, although not stellar, in fact some mottling is suggested with averted vision. To resolve the cluster excellent steady seeing and a larger aperture is required. In a 16 inch telescope some of the outer stars (magnitude 14 to15) emerge with averted vision. This cluster is however a binocular object and is in fact in the American Astronomical League's "Deep Sky Binocular Club's Object List", all of which are visible in 7 × 50 binoculars. You can qualify for a certificate by seeing, in binoculars of course, 60 objects on their list. To find C47 is quite straight forward, just head 3.9° south from Epsilon Delphini, the "tail" of the dolphin.

Object Description

C47 (NGC 6934) is one of the least studied globular clusters in the northern hemisphere and it was not until 1973 that the first colour-magnitude (CM) diagram for it was produced. This was derived by Harris & Racine, the former is of course better known today for his definitive catalogue *Globular Clusters in the Milky Way* (1996/7). The distance to the cluster was determined using both the CM diagram and by studying the periods and luminosity for 31 RR Lyrae variables known in the cluster. The distance to the cluster is about 50,000 light years. More recent research by Piotto et al. (1999) used Hubble Space Telescope observations of the central region of C47 where they found many blue straggler stars (BSS). Using observations of over 8,000 stars they were able to calculate a deeper CM diagram plus the luminosity function. From these, they were able to derive C47's age, which they placed in the range of 13 to 16 billion years. C47 was also a first test object for the Gemini North 8.1 metre telescope in June 1999 and, with adaptive optics, it was able to resolve to the cluster's core.

Galaxy in Cancer

Image

Pedro Re

Database

Name and/or Catalogue Designations:
NGC 2775
H I 2
UGC 4820
PGC 25861

Type of object: Galaxy

Catalogue position for epoch J2000.0
Right ascension: 09h 10m 20.5s
Declination: +7° 02′ 19″

Constellation: Cancer

Object information:
Magnitude: 10.1
Size: 5.0′ × 4.0′
Position angle: 155°
Object classification: Sa
NGC Description: considerably bright, considerably large, very gradually very suddenly
    much brighter towards the middle, resolvable – mottled
Note: in group with NGC2777 and NGC2773

Star atlas chart numbers:
Millennium Star Atlas, Charts 759–760, Volume II
Sky Atlas 2000.0, Chart 13
Uranometria 2000, Chart 187, Volume 1

Finder Chart

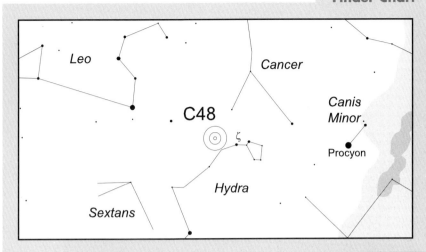

Visual Description

Quite an unusual galaxy and the NGC description of Dreyer that it is "partly resolvable" is pure optimism! At magnitude 10.2 it is not a binocular object but an easy telescope one. An 8 inch (200 mm) telescope reveals it to be quite large, about 5×4 arc-minutes. Detail is definitely lacking (see below for the reasons) but a certain amount can be distinguished, namely a non-stellar brighter core area and a quite distinct fainter outer zone. Deep images reveal a third fainter outer area, which is visually probably out of range of all but the largest telescope – Lew Gamer with his 20 inch Dobsonian reported detecting it. There are two other galaxies in the group, about $\frac{1}{4}$ degree away. NGC2777 to the north-east and NGC2773 to the north-west. They are faint and just visible in a 16 inch at magnitudes 14.2 and 15 respectively. The nearest bright star is Zeta Hydrae (magnitude 3.1). C48 is then 4° ENE.

Object Description

Although classified as a spiral galaxy (Sa) and viewed almost face-on, it has no prominent spiral arms or dark lanes. This is the reason for it being visually rather featureless. What appears to have happened is that this spiral galaxy turned its star making inter-stellar material into stars very early in its life. Proof of this can be found in its spectrum, which is characterised by the presence of absorption lines rather than by emission features more typical of spirals with prominent arms and star forming regions. This is particularly noticeable in the Calcium H and K lines. C48's spectrum is therefore dominated by the light from old cool stars, which themselves are characterised by absorption, not emission lines. This is more common in elliptical or S0 type galaxies. C48 therefore has a smooth, regular appearance with faint spiral arms and little star formation now taking place. A 13.9 magnitude supernova was discovered in C48 by Treffers et al. (1993z) using an automated 760 mm telescope. It was a class Ia type and when first detected was about 4 weeks past maximum.

C49

Image

Buil-Thouvenot

Database

Name and/or Catalogue Designations:
Rosette Nebula
NGC 2237-9

Type of object: Bright Nebula

Catalogue position for epoch J2000.0
Right ascension: 06h 32m 18.0s
Declination: +4° 59' 00"

Constellation: Monoceros

Object information
Magnitude: 5.5
Size: 70.0' × 80.0'
Object classification: emission
NGC Description: pretty bright, very very large, diffuse, part of extremely large nebula ring
   around C50 (NGC 2244)
Note: sometimes referred to as the RMC (Rosette Molecular Complex)

Star atlas chart numbers:
Millennium Star Atlas, Charts 227–228, Volume I
Sky Atlas 2000.0, Chart 12
Uranometria 2000, Chart 227, Volume 1 and 2

Finder Chart

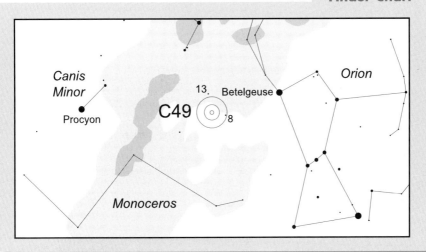

Visual Description

For many, this nebula has often been thought to be too faint to be observable but this is definitely not the case. What does make it difficult is its huge size of well over 1°. The first step to viewing it is to have a field of view of at least double this. This gives sufficient contrast to set it off against the background stars. The telescope needed is a richest field (RFT) type and a 6 or 8 inch (150–200 mm) ideal. Alternatively 20 × 80 binoculars held firm are also adequate but they do have one disadvantage. We are going to need a UHC nebula filter and obviously binoculars will need two of them. Having got our wide field and UHC filter in place all that is needed is to locate the cluster C50 and then, with averted vision, scan for the faintest of glows outside these stars. The brightest knot is to the north-west and there are reports of this part being visible from suburban skies (with a UHC filter of course) but this is perhaps optimistic and really out of town skies should be regarded as essential for C49.

Object Description

The Rosette Molecular Cloud (or Complex), to give it its technical name, is a vast cloud of dust and gas with the open cluster C50 at its centre. This young cluster, formed from the material of the nebula, created the central cavity as radiation and stellar winds from it blew the remaining material outwards. UV radiation from the hottest stars causes the nebula to shine. Research by Schneider et al. (1998) supported this scenario. They found evidence for a "clumpy molecular cloud" exposed to UV radiation producing "Photon Dominated Regions". They also confirmed the outflow of material. At the edge of the nebula, White et al. (1997) observed what are known as "cometary globules". These are low mass molecular clouds with a head-tail appearance and several have infra-red point sources within them. It is thought that star formation is taking place inside them resulting from what is known as "Radiation Driven Implosion".

C50

Open Cluster in Monoceros

Pedro Re

Name and/or Catalogue Designations:
NGC 2244
H VII 2
Collinder 99
Melotte 47

Type of object: Open Cluster

Catalogue position for epoch J2000.0
Right ascension: 06h 32m 24.0s
Declination: +4° 52' 00"

Constellation: Monoceros

Object information:
Magnitude: 4.8
Size: 24.0'
Number of stars: 100 (approx. 34 OB members)
Magnitude of brightest star: 5.88 (12 Monocerotis)
Object classification: II 3 r n:
NGC Description: cluster, beautiful, stars scattered. (12 Mon included)
Notes: within the Rosette Nebula, part of larger Monoceros OB2 association

Star atlas chart numbers:
Millennium Star Atlas, Charts 227–228, Volume I
Sky Atlas 2000.0, Chart 12
Uranometria 2000, Chart 227, Volume 1 and 2

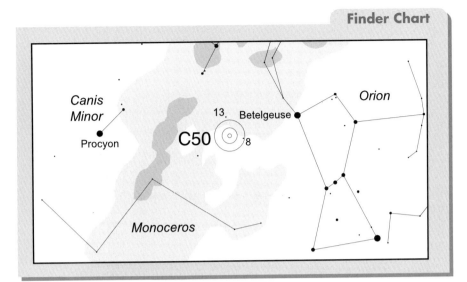

If you have succeeded in finding C49 than you will have already seen the open cluster C50. If you failed then the good news is C50 is very much easier with its brightest stars (about 6 of them to magnitude 8) visible in binoculars. An 8 inch (200 mm) lifts the number of stars visible to around 30. Its brightest star, 12 Monocerotis, does not mark the exact centre, which is located more towards the north-west. 12 Mon is in fact not a cluster member but a foreground object and if you look carefully it is not the blue-white colour of the true members but a touch yellowish. A feature of this scattered cluster is three bright east-west pairs of stars starting at 12 Mon and running to the NNW. The middle pair is actually a pair of pairs. C50 is to be found 9.5° ESE from Betelgeuse (mag. 0.45). Stars 8 & 13 Monocerotis surround it.

We have already learnt (see C49) that C50 is a very young open cluster recently formed at the heart of the Rosette Nebula. It is part of the larger Monoceros OB2 Association and C50 is thought to be located on the edge of the Perseus spiral arm of the Milky Way (Riddle 1972 and Guserva et al. 1985). The distance to C50 is around 5,000 light years. An excellent review of it was published in 1991 by Perez which drew together much of what is known about this cluster. It seems that the most likely scenario for triggering its formation was the collision of two molecular clouds. This formation taking place very rapidly around 3 million years ago (Lynga, 1981) or 4 million years (Ogura/Ishida, 1981). It is interesting that cluster (kinematic) ages can also be estimated on the basis of their radial velocities and total mass. For C50, this age gives excellent agreement at 4 million years. A feature of the cluster is the large number of candidate T Tauri type stars present (several hundred according to Adams et al., 1983). These are believed to be very young proto-stars.

C51

Image

Buil-Thouvenot

Database

Name and/or Catalogue Designations:
IC 1613
UGC 668
PGC 3844

Type of object: Galaxy

Catalogue position for epoch J2000.0
Right ascension: 01h 04m 48.4s
Declination: +2° 07' 10"

Constellation: Cetus

Object information:
Magnitude: 9.2
Size: 11.0' × 9.0'
Position angle: 50°
Object classification: dI (dwarf Irregular)
NGC Description: faint, extremely large
Note: member of Local Group

Star atlas chart numbers:
Millennium Star Atlas, Charts 267–268, Volume I
Sky Atlas 2000.0, Chart 10
Uranometria 2000, Chart 217, Volume 1 and 2

Finder Chart

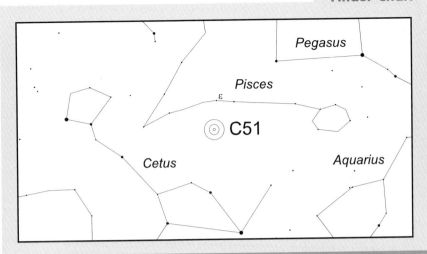

Visual Description

The western part of Cetus is a pretty barren area but there is one object of considerable interest, namely C51 (IC 1613). Interesting because it is a member of the Local Group of Galaxies but very tricky visually as it is a dwarf irregular type. Nevertheless it is quite large (11 × 9 arc-minutes) but spread out so its total magnitude of 9.2 is deceptive. There are reports of it being visible in giant (80 mm) binoculars as a very faint glow but realistically an 8 inch (200) telescope is the minimum, and then really dark skies are needed to provide the necessary contrast. In my experience dark skies are more important than aperture for this object. If you can combine the two then there are reports of brighter areas becoming apparent amongst a dim glow. On the plus side, C51's location can be pinpointed pretty accurately as it lies 12 arc-minutes due south of a 7th magnitude star, easily visible in finderscopes. This star is 5.5° south of Epsilon Piscium.

Object Description

The dwarf irregular galaxy, C51, is a member of the Local Group and has many similarities to C57 (Barnard's Galaxy). Both are about as far away as M31 and somewhat smaller than the Magellanic Clouds. C51 has been extensively studied – I found over 200 research papers referring to it! It appears to be a very old galaxy that is undergoing much star formation at the present time. It does however have, according to Hodge et al., around 77 H-II regions, five of which are complexes of several smaller emission regions. In addition they identified 20 stellar associations (concentrated in the NE and NW quadrants), the core, 43 star-cluster candidates and one relatively certain identification of a dust cloud. An enigmatic X-ray source (IC1613 X-1) apparently emanating from C51 has turned out to be most likely from an unknown background cluster of galaxies (Eskridge, 1995).

C52

Image

Pedro Re

Database

Name and/or Catalogue Designations:
NGC 4697
H I 39
MCG-1-33-10
PGC 43276

Type of object: Galaxy

Catalogue position for epoch J2000.0
Right ascension: 12h 48m 35.9s
Declination: −5° 48′ 02″

Constellation: Virgo

Object information:
Magnitude: 9.2
Size: 6.0′ × 3.8′
Position angle: 70°
Object classification: E4
NGC Description: very bright, large, irregular extended P.A. 45°, small bright nucleus in the middle
Note: dominant member of NGC4697 Group

Star atlas chart numbers:
Millennium Star Atlas, Charts 795–796, Volume II
Sky Atlas 2000.0, Chart 14
Uranometria 2000, Chart 239, Volume 1 and 2

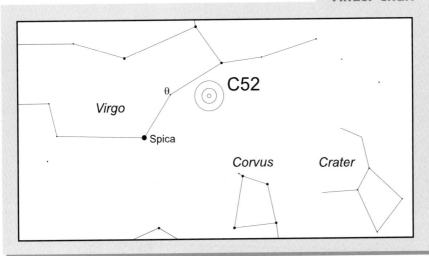

Visual Description

Not far from M104, the Sombrero Hat Galaxy, the classical elliptical galaxy C52 is often passed by. It is true that its appearance is a bit disappointing but it is reasonably bright and one of the few galaxies that stand out in suburban skies. We have no nearby classical elliptical galaxies (other than dwarfs) and this represents an excellent opportunity to study one. An 8 inch (200 mm) reveals an elliptical glow orientated north-east to south west, about 4×3 arc-minutes in size. A 16 inch (400 mm) boosts its detection, the overall size increases by about an arc-minute in each direction. The nucleus becomes more obvious but being an elliptical galaxy it is never going to be spectacular. It is the dominant member of its own group (see below) and for CCD imagers it is worth searching out the adjacent weird spiral NGC4731 and the barred spiral MCG-1-33-3. They are both about 50 arc-minutes away, to the SE and WSW respectively. C52 is located 5.3° west of Theta Virginis (mag. 4.4).

Object Description

C52 is a classical E4 (or E5) galaxy. It is known to be rotating rapidly and this property has been studied by Carter (1987). By comparing it to homogeneous theoretical models marked deviations were detected. This could best be explained by the presence of a weak disc within the galaxy, inclined at an angle of 10 degrees to our line of sight. C52 is the dominant elliptical in a galaxy group of about 18 in number, which includes 2 spirals. It is number 314 in the Local Galaxy Group Catalogue (LGG, 1993) and is one of several elliptical galaxies that are considered standard calibrators for assessing distances. This was achieved by determining the cluster distance from its spiral galaxies, which then could be applied to the dominant elliptical, in this case C52 (Garcia et al, 1995). That distance? Well it would seem to be closer than that catalogued by Tully (1988) and is now believed to be just under 60 million light years away.

C53

Buil-Thouvenot

Database

Name and/or Catalogue Designations:
Spindle Galaxy
NGC 3115
H I 163
MCG-1-26-18
PGC 29265

Type of object: Galaxy

Catalogue position for epoch J2000.0
Right ascension: 10h 05m 14.1s
Declination: −7° 43′ 07″

Constellation: Sextans

Object information:
Magnitude: 8.9
Size: 8.3′ × 3.2′
Position angle: 40°
Object classification: SO
NGC Description: very bright, large, very much extended in position angle 46 degrees, very
   gradually small bright extended nucleus in the middle
Note: its "smooth" appearance has caused it to be mis-classified as an elliptical galaxy

Star atlas chart numbers:
Millennium Star Atlas, Charts 803–804, Volume II
Sky Atlas 2000.0, Chart 13
Uranometria 2000, Chart 279, Volume 2

Finder Chart

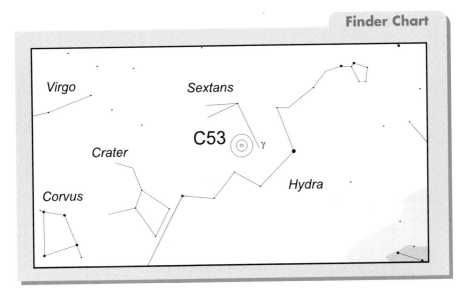

Visual Description

C53, the Spindle Galaxy, is a bright S0 galaxy, i.e. a disc galaxy without significant spiral structure or dust lanes, in the faint constellation of Sextans. It is seen almost exactly edge-on, hence its name. It is bright (mag. 8.9) but, more importantly, it has a high *surface* brightness making it an easy telescope object, even from urban skies. In an 8 inch (200 mm), its elongated shape is clear but it is not as slender as other edge-on galaxies we have already met. It is aligned NE to SW. It gets gradually brighter to the centre with no noticeable steps. Bigger telescopes, e.g. 16 inch, do not show any more detail but the galaxy just gets a bit longer. It has a faint 15th magnitude companion galaxy (PGC29299) 5.5 arc-minutes to its east. C53 was discovered by William Herschel in the 1780's. It lies 3.1° west of, and slightly north of, Gamma Sextantis (mag. 5.0).

Object Description

Thought to be several times bigger than our own Milky Way, galaxy C53 is composed of mainly of old stars. It now contains virtually no gas and very little is going on, apart from the stately orbits of its stars. However, its main claim to fame is that it is one of the best candidates for harbouring a black hole in its midst. Using observations with the Hubble's Faint Object Spectrograph and the Canada-France-Hawaii Telescope, a team of astronomers, led by John Kormendy of the Institute for Astronomy, Honolulu, Hawaii, was able to measure the velocities of stars in the nucleus of NCG 3115. It had been previously shown (Kormendy & Richstone, 1992) that C53 was a candidate for having a massive dark object (MDO) at its centre. These new observations, at unprecedented resolution, showed enormous stellar velocities close to the galactic centre, which could only be explained by the presence of an MDO. These velocities were so high that without the presence of an MDO the "nucleus would explode". So what is that MDO? The new results make it highly likely that it is a supermassive black hole.

C54

Image

Buil-Thouvenot

Database

Name and/or Catalogue Designations:
NGC 2506
H VI 37
Collinder 170
Melotte 80

Type of object: Open Cluster

Catalogue position for epoch J2000.0
Right ascension: 08h 00m 12.0s
Declination: −10° 47′ 00″

Constellation: Monoceros

Object information:
Magnitude: 7.6
Size: 7.0′
Number of stars: 150
Magnitude of brightest star: 10.7 (Hipparcos)
Object classification: I 2 r
NGC Description: cluster, pretty large, very rich, stars 11 to 20 magnitude
Notes: about 150 stars to magnitude 20

Star atlas chart numbers:
Millennium Star Atlas, Charts 833–834, Volume II
Sky Atlas 2000.0, Chart 12
Uranometria 2000, Chart 275, Volume 2

Finder Chart

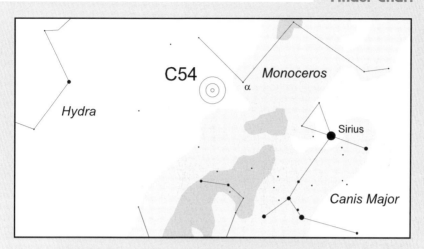

Visual Description

This little observed old southern cluster deserves to be better known. It is not far from Sirius so if that star is visible then it is that time of year to look for C54. It is at its best in a large telescope but it is still worth searching out with a modest one or even with binoculars. It has three stars brighter than magnitude 11 and a small telescope (6 inch–150 mm) shows these and several more, mainly split into eastern and western groups. In a 16 inch telescope the number of stars resolved becomes more impressive (perhaps 50 or so) and, because it is a relatively condensed cluster, it is quite a spectacular sight. It lies in the south-east corner of Monoceros, in a very inconspicuous area. Alpha Monocerotis (4th mag.) is the nearest guide star and is almost 15° due south of Procyon. C54 is then 4.8° ESE from Alpha Monocerotis.

Object Description

Probably the most comprehensive data for this old cluster is that published by Chiu & Van Altena in 1981. They studied 724 stars in the region of C54 to determine cluster membership, which includes many red giants, subgiants, some "blue stragglers" (4 definite) and binary stars. Blue stragglers are old stars that glow with the blue light of young stars, possibly rejuvenated by the merger of two stars. The relative lack of low mass stars was put down to repeated passages of the cluster through the plane of the Milky Way stripping them out. More recently, Twarog et al. (University of Kansas) have determined the colour-magnitude diagram for C54 from which they derive the age for this cluster to be between 2 to 2.3 billion years. However this research was primarily to compare similar stars in the solar neighbourhood with those in the Magellanic Clouds so that a better estimate of the latter's distance could be determined.

Planetary Nebula in Aquarius

Tim Puckett

Name and/or Catalogue Designations:
Saturn Nebula
NGC 7009
H IV 1
PK37-34.1

Type of object: Planetary Nebula

Catalogue position for epoch J2000.0
Right ascension: 21h 04m 13.2s
Declination: −11° 22′ 02″

Constellation: Aquarius

Object information:
Magnitude: 8.3
Size: 28.0″ × 23.0″
Magnitude of central star: 12.0
Object classification: 4(6)
NGC Description: magnificent, very bright, small, elliptical
Notes: named by Lord Rosse using his 6 foot telescope

Star atlas chart numbers:
Millennium Star Atlas, Charts 1335–1336, Volume III
Sky Atlas 2000.0, Chart 17
Uranometria 2000, Chart 300, Volume 2

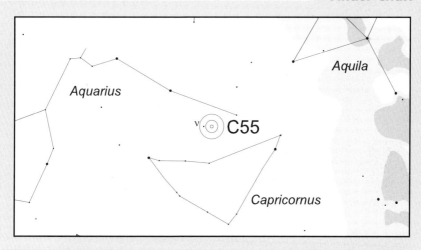

Visual Description

The NGC description of "magnificent" is certainly justified and even a small telescope will be able to clearly see this bright planetary nebula. It got its name, as would be expected, from its similarity to Saturn. Its diameter at 25 arc-seconds is similar to Saturn's disc but it is its "extensions" or ansae on either side, mimicking edge-on rings, that really clinches the similarity. However, these are extremely faint and to see them requires excellent skies and a telescope around 10 to 12 inches in aperture, ideally larger. They are orientated almost East-West. What all telescopes will see however is the bright blue-green disc but the blue central star of 12th magnitude can be elusive. The father of deep-sky observing, Walter Scott Houston, required a 20 inch refractor to see it! To find the nebula sweep west from Nu Aquarii (mag. 4.5) about 1.3°. It helps if a magnification of at least 100× is used as below this C55 appears stellar and insignificant. It could therefore be missed with too low a power.

Object Description

The Saturn Nebula was discovered by William Herschel in 1782 using his 18.7inch f/13 speculum telescope. He observed it again in 1784 with a new 18.7 inch f/13 telescope at powers up to and including 932× – which gives some idea of the optical quality of his instruments! C55 comprises a shell of glowing gas surrounding a hot central star. The nebula was formed by the sudden release of a large portion of the star's mass leaving the blue-hot core of the original star. The intense ultraviolet radiation from which causes the nebula to glow. A recent theory (Soker 1997) is that the peculiar shape of C55 is caused by the central star having an unseen sub-stellar companion. Probably the most detailed image of this nebula was taken by the 3.5 metre WIYN telescope during its commissioning phase. The blue-green colour is due to the presence of double ionised oxygen. C55 is believed to be 3900 light years from Earth and about half a light year across.

Image

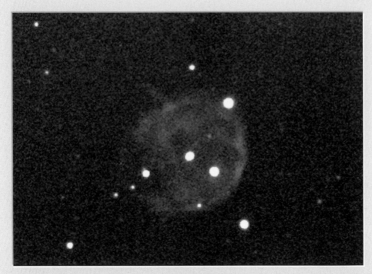

Pedro Re

Database

Name and/or Catalogue Designations:
NGC 246
H V 25
PK 118-74.1

Type of object: Planetary Nebula

Catalogue position for epoch J2000.0
Right ascension: 00h 47m 00.9s
Declination: −11° 52′ 37″

Constellation: Cetus

Object information:
Magnitude: 8.5
Size: 4.0′ × 3.5′
Magnitude of central star: 10.9 (11.7 Hipparcos)
Object classification: 3b
NGC Description: very faint, large, 4 stars in diffuse nebula
Note: galaxy NGC 255 is 25′ to the NNE

Star atlas chart numbers:
Millennium Star Atlas, Charts 315–316, Volume I
Sky Atlas 2000.0, Chart 10
Uranometria 2000, Chart 261, Volume 2

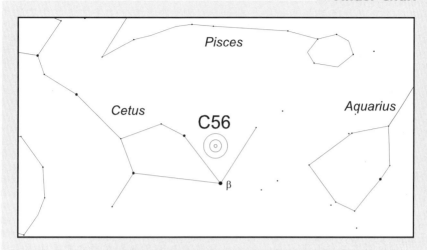

Visual Description

One of the largest of the planetary nebulae (4 × 3 arc-minutes) and one of the most interesting both for its shape and its exotic central star (see below). In a small telescope C56 looks like a small open cluster. Its large size means that its magnitude 8 rating is deceptive and requires a medium sized telescope to reveal its true nature. An 8 inch (200 mm) shows an oval glow with two stars overlaying it, with the NE one being the interesting progenitor star. A fainter one is visible with averted vision. A fourth, easier star is just outside on the NW. Doubling the telescope aperture then it becomes obvious that the mottled appearance is that of a planetary nebula with several brighter and darker areas. The best chance of finding C56 is by locating Phi1 and Phi2 Ceti (both 5th mag.), which are 8° north of Beta Ceti. Then head back 1.25° south.

Object Description

C56 is a high excitation planetary nebula with one of the hottest (>130,000°K) central stars known (Werner and Rauch 1994). This star has a low-amplitude pulsating period of 1500 secs, described as a non-radial g-mode pulsator! It is of the type known as PG 1159 (Bond 1995). These are exotic high mass-loss stars, which are deficient in hydrogen but are helium rich. They are thought to be immediate descendants of Wolf-Rayet stars but are pre-white dwarf. The star in C56 is known to be binary, which may account for the nebula's non-symmetry. C56 has been extensively studied by Reed and Ye Feng, 1997, in the light of H-alpha and O-III. They detected several elements and noticed two small jets coming from the south rim in the O-III image. They were tempted to identify them as FLIERS (fast low-ionisation emission regions – see C22) but the nature of the jets is still unclear and the lack of their symmetric counterparts in the north makes this uncertain.

Image

Steve Lee

Database

Name and/or Catalogue Designations:
Barnard's Galaxy
NGC 6822
MCG-2-50-6
PGC 63616

Type of object: Galaxy

Catalogue position for epoch J2000.0
Right ascension: 19h 44m 57.9s
Declination: −14° 48′ 11″

Constellation: Sagittarius

Object information:
Magnitude: 8.8
Size: 20.0′ × 10.0′
Object classification: IB(s)m
NGC Description: very faint, large, extended, diffuse
Note: IC 1308 and IC 4895 are H-II regions in NGC 6822

Star atlas chart numbers:
Millennium Star Atlas, Charts 1339–1340, Volume III
Sky Atlas 2000.0, Chart 16
Uranometria 2000, Chart 297, Volume 2

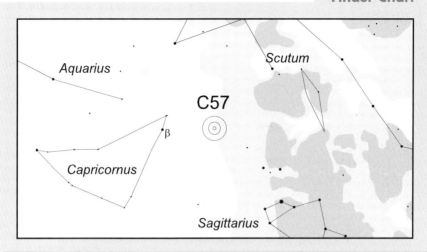

Visual Description

Its fame in cosmological history (see below) is matched by its fame for being visually elusive! In researching visual observations of C57 I found cases where large (over 12 inches–300 mm) telescopes had failed but others where small (4 inches–100 mm) ones had succeeded. Its surface brightness is so low that the first key to seeing it is low power and dark skies. Out of town it can be detected as a hazy patch running roughly east-west – this is the "bar" of this irregular galaxy. The second key is looking in exactly the right place. It is located near the 10th magnitude planetary nebula, NGC6818 (the Little Gem Nebula). This represents the best way to locate C57 but requires higher powers to detect it. Once the nebula is spotted, switch to low power and head 40 arc-minutes SSE. They are both around 8.5° west of Beta Capricorni (mag. 3).

Object Description

The description of C51 referred to the similarities between that galaxy and this, C57, the famous galaxy known as Barnard's Galaxy (not to be confused with "his" star or loop!). C57 is slightly closer to us than C51, at about 1.8 million light years away. Its fame is based on Edwin Hubble's discovery of Cepheid variable stars in C57 (NGC 6822) in 1924, with parallel work in M33 and M31. With this, Hubble settled decisively the fundamental question of the nature of the galaxies, i.e. they were systems external to the Milky Way. Others had *postulated* it (Curtis, Lundmark, and Öpik) but Hubble *proved* it. In contrast to C51, Barnard's galaxy is still actively forming stars and the star formation history has been determined by Gallart et al. (1996), using the 2.5 meter Isaac Newton Telescope. They concluded that C57 most likely begun forming stars at a very early epoch (about 12–15 billion years ago) and that an overall enhancement of the star formation activity has occurred in the last 100–200 million years.

C58

Image

Pedro Re

Database

Name and/or Catalogue Designations:
NGC 2360
H VII 12
Collinder 134
Melotte 64

Type of object: Open Cluster

Catalogue position for epoch J2000.0
Right ascension: 07h 17m 48.0s
Declination: −15° 37′ 00″
Constellation: Canis Major

Object information:
Magnitude: 7.2
Size: 13.0′
Number of stars: 80
Magnitude of brightest star: 10.4
Object classification: II 2 m
NGC Description: cluster, very large, rich, pretty compressed, stars magnitude 9 to 12
Note: at a distance of about 4750 light years

Star atlas chart numbers:
Millennium Star Atlas, Charts 321–322, Volume I
Sky Atlas 2000.0, Chart 12
Uranometria 2000, Chart 274, Volume 2

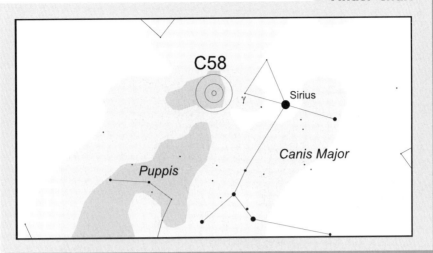

We return to the realm of Sirius. Canis Major has three showy clusters, the better known M41 and the Tau Canis Cluster (C64) plus this little gem, C58 (NGC 2360). It is well within reach of binoculars as a small hazy patch. In an 8 inch (200 mm) telescope about 30–40 stars are resolved in a tight $\frac{1}{4}$ degree swarm. These stars are 10–11th magnitude with the most prominent one on the north-west and some, more imaginative, observers seeing them in lines or rows. In the south of the cluster the stars fade into the background Milky Way. It is located 3.4° east of Gamma Canis Majoris. There is a 4.5 magnitude star 22 arc-minutes west of the cluster.

C58 is a very old open cluster, although not as old as some we have already met, but its estimated age of 1.3 billion years puts it up there with the oldest. It was probably discovered by Caroline Herschel (1750–1848), William's sister, who was an avid astronomical observer and discoverer of comets (she originally found 8 of them) and deep-sky objects ("nebulae"). In both Herschel's catalogue and observing notes, Herschel attributes H VII 12 (C58) to Caroline. On the other hand, in the end-notes at the conclusion of his first catalogue, Herschel notes that Caroline discovered H VII 13 (NGC 2204), but he does not there refer to HVII 12. The most likely explanation (Bryson) is that this was an error of 1 and Caroline did discover C58. In a letter to a friend, Caroline wrote, "William is away, and I am minding the heavens. I have discovered eight new comets and three nebulae never before seen by man... I actually like that he is busy with the Royal Society and his club, for when I finish my other work I can spend all night sweeping the heavens". We amateur observers know what she was getting at!

Planetary Nebula in Hydra

Image

Buil-Thouvenot

Database

Name and/or Catalogue Designations:
Ghost of Jupiter
NGC 3242
H IV 27
PK 261+32.1

Type of object: Planetary Nebula

Catalogue position for epoch J2000.0
Right ascension: 10h 24m 48.6s
Declination: −18° 38′ 14″

Constellation: Hydra

Object information:
Magnitude: 8.6
Size: 40.0″ × 35.0″
Magnitude of central star: 11.0
Object classification: 4(3b)
NGC Description: remarkable, very bright, little extended position angle 147 degrees,
    45 arc-seconds diameter, blue
Notes: very very faint outer shell 24′ × 18′

Star atlas chart numbers:
Millennium Star Atlas, Charts 851–852, Volume II
Sky Atlas 2000.0, Chart 13
Uranometria 2000, Chart 325, Volume 2

Finder Chart

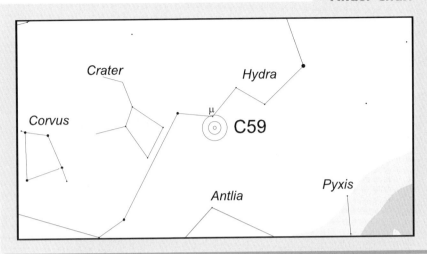

Visual Description

The name Ghost of Jupiter was coined by the noted British observer, Admiral Smyth, in the 19th century and the name has stuck. He noticed the similarity in size to Jupiter itself but that it had a much fainter ghostly glow. Nevertheless it is one of brighter planetary nebula at 8th magnitude and if you can see the Ring Nebula (M57) you should have no trouble with this one. A 6 inch (150 mm) telescope is sufficient to see its outline and colour, either blue or green depending on how your eyes work when viewing faint colours. In larger telescopes (12.5 inch/16 inch) its shape becomes oval with Walter Scott Houston describing it as "eye" shaped – I would agree. It bears magnification well, revealing more detail with the central star easier to spot, although it can be elusive and certainly harder to see than its 11th magnitude would indicate. To locate C59 first find the 4th magnitude star Mu Hydrae, which forms a right angle triangle with Regulus and Spica. The "ghost" is then just under 2° south.

Object Description

It was William Herschel who created the term "planetary Nebula". He could be forgiven for using the term having himself discovered a planet (Uranus). He did realise they were not actually planets but thought they were clouds of gas condensing into stars and planets. We now know that they are the opposite, old stars throwing off material, not young ones just forming. The blue colour of C59 comes from the presence of doubly ionised oxygen but other gases are present in the nebula as well. Its distance is poorly determined but Hajian & Terzian (1995) attempted to derive its distance based on expansion velocities and their determination of 1370 light years is lower than previous estimates. C59 has a very faint outer halo discovered by Bond (1981). This, he thought, must have been ejected from the central star about 10,000 years ago before it went on to eject, just a few thousand years ago, the two visible shells we see today. The central star, according to Kudritzki et al. (1981), is extremely hot with a temperature of 50,000–100,000°K.

C60

Galaxy in Corvus

Buil-Thouvenot

Name and/or Catalogue Designations:
The Antennae (with C61)
NGC 4038
H IV 28
Arp 244
PGC 37967

Type of object: Galaxy

Catalogue position for epoch J2000.0
Right ascension: 12h 01m 52.9s
Declination: −18° 51′ 54″

Constellation: Corvus

Object information:
Magnitude: 10.3
Size: 2.6′ × 1.8′
Position angle: 80°
Object classification: Sp
NGC Description: pretty bright, considerably large, round, very gradually brighter in the middle
Notes: alternative names − Ring-tail or Rat-tail Galaxy

Star atlas chart numbers:
Millennium Star Atlas, Charts 845–846, Volume II
Sky Atlas 2,000.0, Chart 14
Uranometria 2,000, Chart 328, Volume 2

Finder Chart

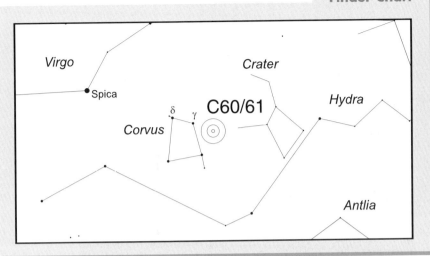

Visual Description C60 & C61

After M51, these objects are probably the easiest interacting galaxies to observe. Their double nature is clear in only an 8 inch (200 mm) telescope with C60 (NGC4038) obviously the brighter (C61 is the southern part). A 12.5 inch (320 mm) will show the connecting bridge, although from Britain they are low in the sky so pristine conditions are required. I was able to see them clearly after heavy rain, which had cleaned the atmosphere perfectly! To locate them, follow the line from Delta Corvi to Gamma Corvi and on an equal amount. These two stars and C60/1 form an equally spaced triple.

Object Description C60 & C61

Formerly known unofficially as the Ring-tail or Rat-tail Galaxies, these interacting galaxies are now generally known as the Antennae. This is, as one would expect, because they have a pair of long tails of luminous matter, which resemble an insect's antennae, formed by the encounter. The tails are thought to originate when two similar galaxies travelling in elliptical orbits pass partially through each other at their closest approach. The result of the gravitational interaction is that material escapes and is spread out in long streamers or tails. Computer simulations are able to mimic this precisely.

However, it was new high-resolution colour images of these galaxies, taken with the Hubble Space Telescope, which brought them to the attention of the general public when NASA published them. The accompanying press release described them as a "fireworks show". The Hubble observations of the Antennae galaxies,

(continued in C61)

C61

Image

Buil-Thouvenot

Database

Name and/or Catalogue Designations:
The Antennae (with C60)
NGC 4039
H II 282
Arp 244
PGC 37969

Type of object: Galaxy

Catalogue position for epoch J2000.0
Right ascension: 12h 01m 53.9s
Declination: −18° 53′ 06″

Constellation: Corvus

Object information:
Magnitude: 10.6
Size: 3.2′ × 2.2′
Position angle: 130°
Object classification: S(B)p
NGC Description: pretty faint, pretty large
Note: alternative name − Ring-tail or Rat-tail Galaxy

Star atlas chart numbers:
Millennium Star Atlas, Charts 845–846, Volume II
Sky Atlas 2000.0, Chart 14
Uranometria 2000, Chart 328, Volume 2

(continued from C60)
together with others of similar colliding galaxies, were conducted by Whitmore (STScI) and co-investigators Francois Schweizer and Bryan Miller (Department of Terrestrial Magnetism, Carnegie Institution of Washington), and Michael Fall and Claus Leitherer (STScI) over several years. However, the C60/C61 combination is probably the best for studying the formation of stars and star clusters, since it is the nearest and youngest example of a pair of colliding galaxies. They are relatively nearby at a distance of 63 million light years. The NASA team's research has enabled them to put together a chronological sequence of how colliding galaxies evolve. This will help us address one of the fundamental questions in astronomy: why some galaxies are spirals while others are elliptical in shape.

Hubble's high resolution and sensitivity allowed the team to uncover, within the Antennae, over 1,000 exceptionally bright young star clusters, sometimes called super star clusters. The brightest of these clusters contain around one million stars. Ground-based telescopes were only able to see the most prominent of these, and even then they were not able to show that the clusters were extremely compact, no larger than ordinary globular clusters. The results threw up several new findings:-

- The sheer number of these young star clusters was far greater than imagined;

- Unlike those belonging to the Milky Way, globular star clusters are not necessarily the relics of the earliest generations of stars formed in a galaxy, as once commonly thought, but may also provide evidence of more recent collisions. In the antennae the youngest clusters are only a few millions years old;

- The "seeds" for star clusters appear to be huge clouds (tens to hundreds of light-years across) of cold hydrogen gas, called giant molecular clouds. These are squeezed by the surrounding hot gas, which was heated during the collision, and then they collapse under their own gravity. Like a string of "firecrackers" being ignited by the collision, these reservoirs of gas light up in a great burst of star formation;

- The ages of the resulting clusters provide a clock for estimating the age of a collision. This offers an unprecedented opportunity for understanding, step-by-step, the complex sequence of events that takes place during a collision, and possibly even the evolution of spiral galaxies into elliptical galaxies.

C62

Galaxy in Cetus

Image

Steve Lee

Database

Name and/or Catalogue Designations:
NGC 247
H V 20
IRAS 00446–2101
PGC 2758

Type of object: Galaxy

Catalogue position for epoch J2000.0
Right ascension: 00h 47m 08.7s
Declination: −20° 45′ 38″

Constellation: Cetus

Object information:
Magnitude: 9.1
Size: 20.0′ × 7.0′
Position angle: 174°
Object classification: SAB(s)d (formerly Sc)
NGC Description: faint, extremely large, very much extended P.A. 172°
Notes: part of the Sculptor Group

Star atlas chart numbers:
Millennium Star Atlas, Charts 339–340, Volume I
Sky Atlas 2000.0, Chart 18
Uranometria 2000, Chart 306, Volume 2

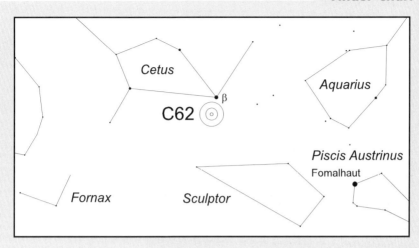

With a magnitude (total integrated) of 9.1 and a size of almost 1/3rd of a degree long one could be forgiven for imagining this galaxy was going to be easy to see! In a medium sized telescope (16 inches – 400 mm) C62 can just be detected as an elongated north–south smudge. Even the nucleus is faint, although it does brighten a bit from the rest of the galaxy. The southern part of the galaxy is the brightest (or should it be the least faint!) and the northern end quite difficult requiring averted vision. Viewing from Britain does not do it justice but when better located, it is easily visible in an 8 inch, at low power. It is a simple star-hop from the 2nd magnitude star Beta Ceti. Head just under 3° SSE. There are a couple of 5th magnitude stars 1° south of C62.

C62 is a nearby galaxy seen from a highly inclined angle, not quite edge-on so that it can be seen to be a fairly open spiral with a hint of a bar. This has lead to its reclassification from Sc to an SAB type i.e. a spiral with a weak bar. It is part of the Sculptor Group of Galaxies (also known as the South Polar Group), which is the closest group to the Local Group, at only about 10 million light years distance. The main group members are NGC 55 (C72), NGC 247(C62), NGC 253 (C65), NGC 300 (C70), NGC 7793 and SDIG (Sculptor Dwarf Irregular Galaxy). However, some very recent research (Catanzarite et al. in preparation at the time of writing) has raised some awkward questions. This concerned their discovery of 9 Cepheids and the subsequent measurement of their periods and luminosities. Cepheid variables are one of the most important "standard candles" for determining distances. Their results showed that C62 was much further away than previously assumed (almost twice as far away as cluster member NGC300) and is now estimated to be 13 million light years distant.

Planetary Nebula in Aquarius

Image

Tim Puckett

Database

Name and/or Catalogue Designations:
Helix Nebula
NGC 7293
PK36–57.1

Type of object: Planetary Nebula

Catalogue position for epoch J2000.0
Right ascension: 22h 29m 37.6s
Declination: −20° 47′ 37″

Constellation: Aquarius

Object information:
Magnitude: 6.3
Size: 16.0′ × 12.0′
Magnitude of central star: 13.5
Object classification: 4(3)
NGC Description: remarkable, pretty faint, very large, extended or binuclear
Note: sometimes referred to as the Sunflower Nebula

Star atlas chart numbers:
Millennium Star Atlas, Charts 1355–1356, Volume III
Sky Atlas 2000.0, Chart 23
Uranometria 2000, Chart 347, Volume 2

Finder Chart

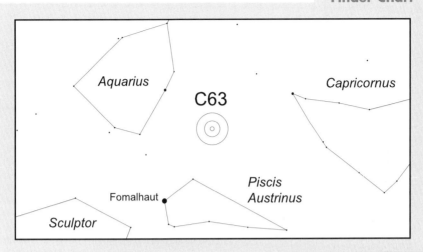

Visual Description

Despite being probably the closest planetary nebula to us C63 is remarkably difficult to see. Its sheer size (nearly as large as the full Moon in photographs) and its extremely low surface brightness means its light is too spread out to produce much contrast against the sky. However modern technology can come to our aid. This is one object where nebula filters really do improve the view. An 8 inch (200 mm) at low power with a good nebula filter (such as an O-III) will magically make it spring into view. Larger telescopes will show the central hole and a 10 inch the central blue star. Don't expect to see the red colour, so prominent in photographs – even in a 16 inch telescope there is only a greeny-grey tint. It is located about 1/3rd the way from Upsilon Aquarii towards 57 Aquarii, a pair of 5th magnitude stars.

Object Description

Called the Helix because of because of it its likeness to a coiled spring, its distance has recently been revised placing it only 450 light years away, virtually next door on a galactic scale. According to Dyson et al. (1989), the progenitor star was probably a red giant, which evolved through a super-wind (low velocity but high mass-loss) during which the bulk of the nebula was formed. After the super-wind, a high velocity low mass-loss wind lasted for around 10,000 years. The exposed core of the star was then at a temperature of around 30,000°K, which was hot enough to ionise the nebula. The star is now past the fast wind stage and its temperature is dropping rapidly. Papamastorakis et al. (1995) detected an extremely faint outer halo, so the progenitor star presumably went through more than one mass-loss phase. Research by O'Dell and Handron (1996), using detailed images taken with the Hubble Space Telescope, has revealed radial streamers, dubbed cometary knots because their glowing heads and gossamer tails resemble comets. Each gaseous head is at least twice the size of our solar system. The most probable theory for their formation is that they are the result of the collision of the different wind phases, the later faster wind ploughing into the slower original(s).

Image

Buil-Thouvenot

Database

Name and/or Catalogue Designations:
Tau Canis Major Cluster
NGC 2362
H VII 17
Melotte 65
Collinder 136

Type of object: Open Cluster

Catalogue position for epoch J2000.0
Right ascension: 07h 18m 48.0s
Declination: −24° 57′ 00″

Constellation: Canis Major

Object information:
Magnitude: 4.1
Size: 8.0′
Number of stars: 60
Magnitude of brightest star: 4.4
Object classification: I 3 p n
NGC Description: cluster, pretty large, rich
Note: 40 stars to 13 mag. surrounding Tau CMA, distance about 5000 light years,
   discovered by Hodierna (published 1654)

Star atlas chart numbers:
Millennium Star Atlas, Charts 345–346, Volume I
Sky Atlas 2000.0, Chart 19
Uranometria 2000, Chart 319, Volume 2

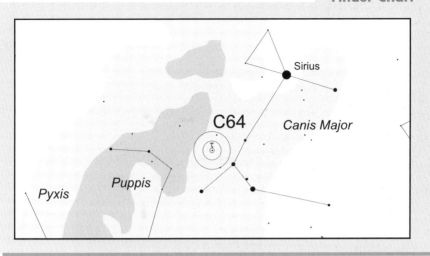

Visual Description

A fabulous open cluster when viewed with any optical aid, from binoculars to big Dobsonians. It surrounds the 4.4 magnitude star Tau Canis Major, which lends the cluster its name. Having a bright star in the cluster adds a real touch of class! Binoculars show Tau CM and an unresolved haze. In a 6 inch (150 mm) telescope the haze reveals itself to be a tight swarm of 25 stars, spread around Tau CM in a circle. In a bigger telescope it just gets better with about 50 stars visible in a 16 inch (more if the cluster is high overhead in your location) and Tau, which is very white, has two bluish companions. Tau CM is located in the "tail" of Canis Major, 11.5° south–east of Sirius

Object Description

C64 is the youngest known open cluster in the Milky Way with an age of between 1 to 2 million years. Much of our information about clusters is derived from Hertzsprung–Russell (H–R) Diagrams. The key is the "Main Sequence (MS) Turnoff". This is at the highest mass star still on the MS. The MS lifetime of this star is the age of the cluster. In C64 however there is no "turnoff" – its highest mass stars are still on MS indicating extreme youth. However, the main interest in C64 is Tau CM, a massive O-type star. It is probably a cluster member but its radial velocity does differ slightly from other members. Using new Hipparcos data van Leewen & van Genderen found that Tau was in fact a quadruple star system. Tau had previously been known to comprise a visual double with one of these a spectroscopic binary. One of the latter components is now believed to be a very close binary as well! They postulate that it probably formed from two binaries in which, due to interactions, one binary became closer, the other wider. A quadruple system such as this is, in general, unstable.

C65

Image

Steve Lee

Database

Name and/or Catalogue Designations:
Sculptor Galaxy or Silver Coin Galaxy
NGC 253
H V 1
PGC 2789

Type of object: Galaxy

Catalogue position for epoch J2000.0
Right ascension: 00h 47m 33.1s
Declination: −25° 17′ 18″

Constellation: Sculptor

Object information:
Magnitude: 7.2
Size: 25.0′ × 7.0′
Position angle: 52°
Object classification: Scp or SB
NGC Description: very remarkable, very very bright, very very large, very much extended
    P.A. 54°, gradually brighter towards the middle
Note: Allan Sandage "arms are more defined by dust than stars".
NGC 288 (globular cluster) 2° to SE

Star atlas chart numbers:
Millennium Star Atlas, Charts 363–364, Volume I
Sky Atlas 2000.0, Chart 18
Uranometria 2000, Chart 306, Volume 2

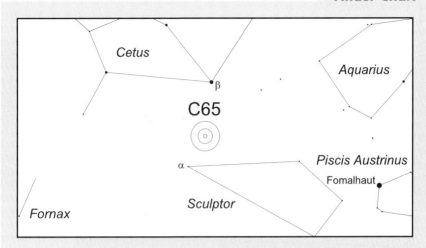

This is the Southern Hemisphere's answer to the Andromeda galaxy being its biggest and brightest. Best viewed from well south of Britain it was nevertheless discovered from Britain by Caroline Herschel (see C58). It is brilliant in any telescope (or binoculars) being a full $\frac{1}{2}°$ long with a high surface brightness. Even a lowly 60 mm refractor will reveal its shape. It is orientated north–east to south–west and is roughly 3 times longer than it is wide. The bigger the telescope however, the better the view. In an 8 inch the size becomes huge and with averted vision the spiral arms are hinted at – the southern one being a touch brighter. For northern observers head 7.3° south from Beta Ceti, whilst southern ones head 4.75° north–west from Alpha Sculptoris. Either way it cannot be missed!

We have already referred to the Sculptor Group (see C62). The dominant member of this group is C65 (NGC253), which is similar to M31 in mass. However it exhibits numerous peculiar features including a nuclear ring, gas outflow, a bar (hence class SAB) and a warp. As if that wasn't enough, it is one of the dustiest galaxies known and it currently has a very high star-forming rate in the region of its nucleus. This central core area has come under the scrutiny of the Hubble Space Telescope (Watson et al.). These images showed four "super star clusters" and their properties are astounding. They contain hundreds of luminous stars and their total masses are about $\frac{1}{2}$ million times that of the Sun. However, all of these stars are packed into a size of only about 5 light years across, which is about the distance from the Sun to the its nearest neighbouring star. These clusters are much brighter and more massive than other known star clusters, and to signify this distinction they have been named "super star clusters".

C66

Image

Buil-Thouvenot

Database

Name and/or Catalogue Designations:
NGC 5694
H II 196

Type of object: Globular Cluster

Catalogue position for epoch J2000.0
Right ascension: 14h 39m 36.5s
Declination: −26° 32′ 18″

Constellation: Hydra

Object information
Magnitude: 10.2
Size: 3.6′
Object classification: 7
NGC Description: considerably bright, considerably small, round, pretty suddenly brighter towards the middle, star 9.5 magnitude south preceding
Note: discovered by William Herschel 1784, identified as a globular cluster by Lampland/Tombaugh in 1932.
Distance from the Sun: 33.9 kpc.
Distance from the galactic centre: 28.3 kpc.

Star atlas chart numbers:
Millennium Star Atlas, Charts 863–864, Volume II
Sky Atlas 2000.0, Chart 21
Uranometria 2000 Chart 332, Volume 2

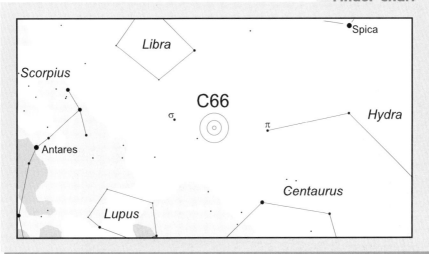

This distant globular cluster is not one of the more spectacular and I guess having seen it once most will pass on to the next Caldwell object! With its brightest stars around magnitude 16.5 it is not going to resolve itself into sparkling members in any ordinary telescope. At low power and in a small telescope it could easily be missed as it looks like a 10th or 11th magnitude star. In an 8 inch (200 mm), at medium power, a circular haze with a somewhat brighter middle is evident, perhaps about 2 to 3 arc-minutes in diameter with averted vision. Bigger telescopes (16 inch) just increase its size slightly, to about 4 arc-minutes and some "mottling" becomes detectable. Steve Coe, observing with a 17.5 inch telescope under excellent skies, thought he could see 6 stars superimposed on C66. However, it is a very difficult object from Britain where just glimpsing it is an achievement and a computerised telescope is all but essential! C66 is located at the tip of Hydra's tail, head either 5.5°WSW from Sigma Libræ or 7.5° due east from Pi Hydræ (both mag. 3.25). Look for a 7th magnitude star 11 arc-minutes to its south.

C66 is a Milky Way globular cluster but, because it is located on the other side of our galaxy to us, it is quite remote. Most of the information regarding the cluster is from "Deep photometry of NGC 5694 and Terzan 8" (Ortolani et al., 1990). They found it to be metal poor (i.e. elements higher than helium) and have a blue horizontal branch, a very high radial velocity and distance of about 90,000 light years from the galactic centre or about 110,000 from us. That high radial velocity has been remarked on before (Burnham) and it has been postulated that C66 has somehow been accelerated to galactic escape velocity. More recent research (Konig & Fahlman, 1995) has concerned enigmatic bright blue quasi-stellar knots very close to the cluster core. The photometry indicated that these are unlikely to be chance super-positions of bright stars. They seem to be part of a stellar population modified by the extreme surroundings in the cluster core.

Tim Puckett

Database

Name and/or Catalogue Designations:
NGC 1097
H V 48
ESO 416–20
ARP 77
PGC 10488

Type of object: Galaxy

Catalogue position for epoch J2000.0
Right ascension: 02h 46m 18.9s
Declination: −30° 16′ 21″

Constellation: Fornax

Object information:
Magnitude: 9.5
Size: 9.3′ × 6.3′
Position angle: 130°
Object classification: SBb
NGC Description: very bright, large, very much extended P.A 151°, very much brighter nucleus in the middle
Note: LINER/Seyfert galaxy, companion galaxy NGC1097A is 5′ to the NW (see image)

Star atlas chart numbers:
Millennium Star Atlas, Charts 381–382, Volume I
Sky Atlas 2000.0, Chart 18
Uranometria 2000, Chart 354, Volume 2

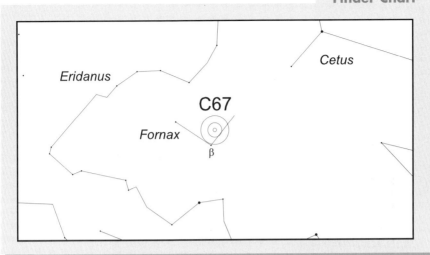

Despite being face-on this is a relatively bright galaxy (magnitude 9) that is easy to see and bears magnification well. It is one of the easiest in which to visually spot a bar. The arms emanating from the ends of the bar are much harder to see and realistically are only in the range of big Dobsonians (16 inches plus). However an 8 inch (200 mm) telescope will reveal the galaxy's core and an elliptical glow orientated SE to NW. This is in fact the bar. Its companion galaxy, located about 5 arc-minutes north–west, is easy despite being only magnitude 13.6. C67 is not an easy object to star-hop to, but using a Telrad finder first centre Beta Fornacis (4th mag.). Then just outside the 4° diameter ring (i.e. 2° away), in the direction NNW from Beta, C67 will be found.

Optically C67 is the classic barred spiral, viewed almost face-on. However, it does have an Active Galactic Nucleus (AGN), which was previously classed as a LINER (see C29) but now a Seyfert Type 1. It also has 4 optical jets extending from it. Wehrle et al. (1997) deduced that because of the blue colour of these jets, the lack of radio emission from them and their morphology, they were most likely to be tidal streamers formed during encounters with its companion, NGC 1097A. However, perhaps the most surprising discovery in C67 is the rapid changes observed in the emission from its nucleus by Storchi–Bergmann et al. (1997). Using the 4 metre Blanco Telescope, they compared Hydrogen Balmer-line spectra of the nucleus taken in 1991, 1994 and 1996. These showed dramatic changes with the last observation changing to a new double-peak profile with the blue peak higher than the red, the opposite of the 1991 observations! These line profiles are well produced in theoretical models for gas rotating at relativistic speeds in a Keplerian accretion disc. The most likely source of the sudden appearance of the double-peak profile is thought to be due to the sudden tidal disruption of a star falling into a black hole. The distance to C67 is, according to Tully, 47 million light years.

Bright Nebula in Corona Australis

Image

Steve Lee

Database

Name and/or Catalogue Designations:
R Coronae Australis Nebula
NGC 6729
Cederblad 165c

Type of object: Bright Nebula

Catalogue position for epoch J2000.0
Right ascension: 19h 01m 48.0s
Declination: −36° 58′ 00″

Constellation: Corona Australis

Object information
Magnitude: 9.7
Size: 1.0′
Object classification: emission/reflection
NGC Description: variable star (from mag. 11…) with nebula
Note: contains stars R CrA (variable 10.0–14.4) and T CrA (variable 11.7–14.3). Globular
    cluster NGC6723 is 1/2° to NW   .

Star atlas chart numbers:
Millennium Star Atlas, Charts 1433–1434, Volume III
Sky Atlas 2000.0, Chart 22
Uranometria 2000 Chart 379, Volume 2

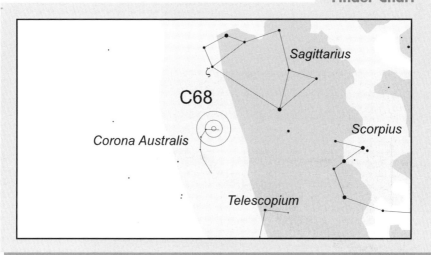

Visual Description

C68 is probably best known because of its availability as one of David Malin's (AAT) many spectacular colour prints. It has many similarities to C46 (Hubble's Variable Nebula) being also variable, comet-like and having a star at its apex. This star is known as R CrA. It is the cause of the variable brightness and is generally itself around 12th magnitude, so normally an 8 inch telescope is about the minimum for a reasonably clear view. It can however, because it varies, be invisible in bigger telescopes or visible in smaller ones! To the north–west is a pair of brighter stars and a nebula (NGC6726–7). For C68 itself, a 12 inch (300 mm) telescope is required for a good view but even so it is small (1 arc-minute across) and it tails away to the south-east towards a fainter star (the associated variable T CrA). A nebula filter is a hindrance rather than a help, indicating that it is mostly a reflection nebula rather than an emission type. To find C68, head 7° south from Zeta Sagittari (bottom of the "teapot") to globular cluster NGC6723. C68 is then 1/2° to the south–east.

Object Description

This predominantly reflection nebula is the result of several bright stars caught up in a large, dusty cloud. However, the dark cloud is a complicated active star forming nebula, though most of the action is hidden from view. R CrA is an emission-line A-type star located near the apex of the comet-like reflection nebula with the young F0e star, T CrA, located near the southern edge of the nebula. NGC 6726/7, to the north–west of R CrA, includes the young B- type, emission-line variable TY CrA. Millimetre-wave observations by Loren(1979) showed that the dominant heating source for this region is near TY CrA. The whole complex is known as the "R CrA Association". Far-IR studies (Wilking et al., 1985) have identified star formation in the association with around 50 embedded stars present.

C69

Image

Steve Lee

Database

Name and/or Catalogue Designations:
Bug Nebula or Bipolar Nebula
NGC 6302
PK349+1.1
Henize 2–204

Type of object: Planetary Nebula

Catalogue position for epoch J2000.0
Right ascension: 17h 13m 44.1s
Declination: −37° 06' 10"

Constellation: Scorpius

Object information:
Magnitude: 12.8
Size: 2.0' × 1.0'
Object classification: 6
NGC Description: pretty bright, extended, preceding and following (the star)
Note: magnitude according to Hubble Guide Star Catalogue = 7.12 (non-star)

Star atlas chart numbers:
Millennium Star Atlas, Charts 1439–1440, Volume III
Sky Atlas 2000.0, Chart 22
Uranometria 2000, Chart 376, Volume 2

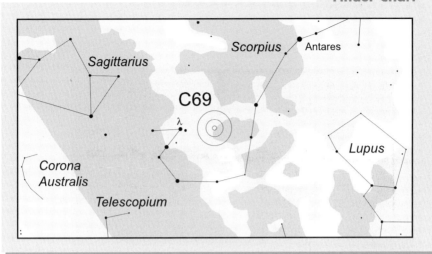

Visual Description

The Bug Nebula, C69, is much brighter than its often quoted *photographic* magnitude of 13. Its centre is more typically 10th magnitude and therefore in reach of even the smallest telescope. It is of the bipolar type of planetary nebula seen sideways-on. Its appearance is of an extended butterfly or figure 8 shape. The long axis is orientated east–west and requires a medium sized telescope to detect its 2 arc-minute length and central narrowing. There are several reported observations of the central star but this is not actually the case (see below). What these observers were seeing is the bright central region of the nebula, which does appear star-like in an amateur sized telescope. The declination of this object places it too far south for observation from Britain but ideal from the now popular astronomy holiday centres in Portugal. It is located 4° due west of Lambda Scorpii (mag. 1.6).

Object Description

As already mentioned, C69 is a classic bipolar nebula and hardly "planetary" looking at all. In fact, nearly 70 percent of all planetary nebula are of this type. The two areas of gas on each side are being ejected from a central star, late in its life, at an extremely high rate. This central star has never been seen, partly because it is faint, but mainly because there is an obscuring dust lane running right across the centre of the nebula. The central star has been studied by Pottasch et al. (1996), although of course not directly, but by the spectrum emitted by the nebula. Their conclusion was that C69 is one of the brightest in the galaxy and that the observations can be best explained by the presence of a very hot central star (about 380,000K), very close to the white dwarf stage. This is still somewhat lower than the temperature proposed by Ashley & Hyland (1988) but still enormous. Its distance, derived from VLA observations (Gomez et al., 1993), is of the order of 5000 light years.

Galaxy in Sculptor

Steve Lee

Database

Name and/or Catalogue Designations:
NGC 300
PGC 3238
ESO 295–20
IRAS00523–3756

Type of object: Galaxy

Catalogue position for epoch J2000.0
Right ascension: 00h 54m 53.8s
Declination: −37° 40′ 57″

Constellation: Sculptor

Object information:
Magnitude: 8.1
Size: 20.0′ × 15.0′
Position angle: 111°
Object classification: Sd
NGC Description: pretty bright, very large, very much irregularly extended, very gradually pretty much brighter towards the middle
Note: Dreyer's notes to the NGC – "A complex object with several nuclei."

Star atlas chart numbers:
Millennium Star Atlas, Charts 407–408, Volume I
Sky Atlas 2000.0, Chart 18
Uranometria 2000, Chart 351, Volume 2

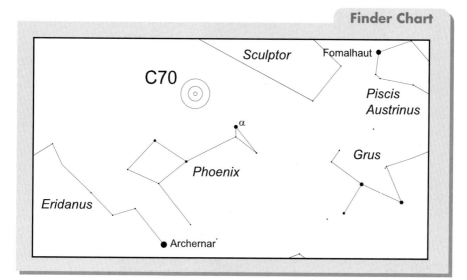

Finder Chart

Visual Description

The best way to describe C70 is as the southern equivalent of M33, the Pinwheel Galaxy in Triangulum. There are many similarities, both are face-on, both have small bright centres and have predominantly blue loose spiral arms. The similarities extend to their visual appearance too and C70 is difficult at first because it is so extended with a low surface brightness. Once spotted in an 8 inch (200 mm) telescope its core is obvious and the haze of its spiral arms (they are not defined at this aperture) can be detected extending over an area a little larger than half of a full Moon. In large binoculars (20 × 80) it looks like a tiny cloud. Being located close to the South Galactic Pole, there are not many bright stars around to point the way – even Alpha Sculptoris is only magnitude 4.3. Instead try heading 7.1° NE from Alpha Phoenicis (mag. 2.4).

Object Description

The similarity to M33 has already been referred to but there are also some differences. C70 has about twice as much of its mass in the form of cold hydrogen and is, as a result, intrinsically only half as bright. The spiral arms too are not quite as dominant. This galaxy is another of those southern spectacular objects photographed with the Anglo-Australian Telescope by the astro-photographic wizard, David Malin. His picture reveals C70 in all its glory with a yellowish haze, from old stars, around the nucleus and beautiful spiral arms containing many hot young blue stars and several prominent pink/red emission nebulae. The galaxy is part of the Sculptor group and its distance, and therefore that of the group, has been the subject of much research. 18 Cepheid variable stars were discovered by Graham (1984) in C70 and his results have been added to (or corrected!) over the years by Madore (1987), Walker (1988), Visanathan (1987) and Freedman (1991) These latest results suggest a distance of 7 million light years.

C71

Image

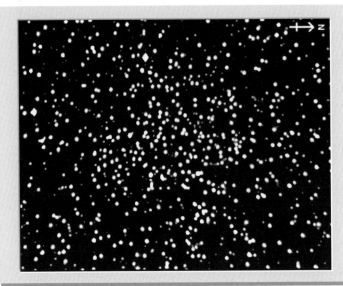

Buil-Thouvenot

Database

Name and/or Catalogue Designations:
NGC 2477
Melotte 78
Collinder 165

Type of object: Open Cluster

Catalogue position for epoch J2000.0
Right ascension: 07h 52m 18.0s
Declination: −38° 33′ 00″

Constellation: Puppis

Object information:
Magnitude: 5.8
Size: 27.0′
Number of stars: 200
Magnitude of brightest star: 12.0
Object classification: I 2 r :b
NGC Description: remarkable, cluster, bright, rich, large, little compressed, star
   12th magnitude
Note: 300 stars to 12th magnitude.
Shapley − "richest of the open clusters"

Star atlas chart numbers:
Millennium Star Atlas, Charts 389–390, Volume I
Sky Atlas 2000.0, Chart 19
Uranometria 2000, Chart 362, Volume 2

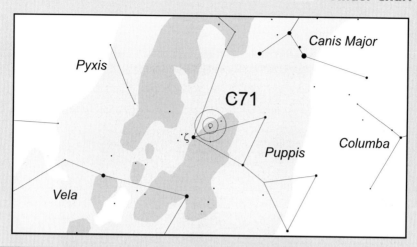

One of the finest binocular objects in the Caldwell Catalogue! Together with its neighbour NGC2451, only 1.5° away, they form a beautiful pair of open clusters in 20 × 80 binoculars. In these, NGC2451 looks the classier with brighter stars (one red) and C71 appearing as a circular haze, just a bit smaller than the full moon. But in a telescope the full majesty of C71 becomes apparent and it easily surpasses its neighbour. In a 6 inch (150 mm), the haze transforms itself into a rich myriad of glittering stars, perhaps 50 resolved with the general haze of others providing a backdrop. The brightest stars are 10th magnitude. In a 13 inch (330 mm) telescope, Steve Coe estimated that 180 stars were visible. Common to all sizes of telescope is a fairly even spread of stars but with several alignments (chains) and dark lanes present. It is situated about 2.5°NW from the star Zeta Puppis (2.25 mag.) with a 4.5 magnitude star (b Pup) lying just to its south–east.

This open cluster was discovered by Lacaille on his 2-year journey to the Cape of Good Hope in 1751–52. It was subsequently observed by Dunlop and entered into his 1827 catalogue. It is such a rich cluster that Shapley was unsure as to whether it was a dense open cluster or a loose globular. There are in total about 300 stars in C71 although many are probably not true cluster members but just lie on the same line of sight. Lohmann (1979) studied this cluster, as well as M36 and M38, finding C71 to be 5 times more massive than M36 and twice that of M38. More recent research by Hippel et al. (1994) and Kassis et al. (1997) has attempted to determine the age of this old cluster. The former group used HST observations of white dwarfs, the latter derived colour magnitude diagrams from deep CCD images. Their conclusions were similar, with a relatively old age of 1–1.2 billion years.

C72

Galaxy in Sculptor

Image

Steve Lee

Database

Name and/or Catalogue Designations:
NGC 55
ESO 293–50
PGC 1014
Dunlop 507

Type of object: Galaxy

Catalogue position for epoch J2000.0
Right ascension: 00h 15m 08.5s
Declination: −39° 13′ 13″

Constellation: Sculptor

Object information:
Magnitude: 7.9
Size: 25.0′ × 4.0′
Position angle: 108°
Object classification: SBp
NGC Description: very bright, very large, very much extended, triple nucleus
Note: member of Sculptor Galaxy Group

Star atlas chart numbers:
Millennium Star Atlas, Charts 429–430, Volume I
Sky Atlas 2000.0, Chart 18
Uranometria 2000, Chart 350, Volume 2

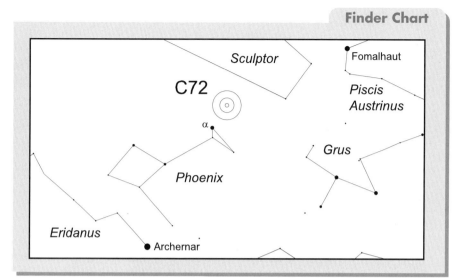

Visual Description

One of the brightest galaxies in the southern hemisphere (C65 is brighter) but never seen at its best from southern Europe or the USA. Discovered by James Dunlop from Paramatta, New South Wales (1827) using a 9-inch telescope. He described it as "a beautiful long nebula, about 25 arc-minutes in length…". He was right and its closeness to the Milky Way enables even amateur telescopes to show interesting detail. Its brightest stars are magnitude 18.5 so they are well within range of amateur CCD imagers! Visually 20×80 binoculars reveal a spindle type haze whilst an 8 inch shows the full extent of its long slender profile. It extends over a 1/3rd degree long, running approximately east-west, with an off-centre brightening towards the west. If you have access to a larger telescope then this is one galaxy where a nebula filter will help, showing several of the brighter H-II regions. It is located in southern Sculptor, 3.7° NNW of Alpha Phoenicis.

Object Description

Sometimes classed as a type I Irregular galaxy, C72 is now generally accepted to be a barred spiral (SB). It is highly inclined (about 9 degrees from edge-on) and is morphologically similar to the Large Magellanic Cloud, but is slightly more luminous. It has numerous bright H-II regions but more interestingly it also has several large filamentary and bubble-like features, extending well above the main body of the galaxy into the halo. These features have been studied by Ferguson et al. (1996). They identified candidate "chimneys'" extending out of the disc, which could be the conduits into the halo for hot gas from around disc star-forming regions, and could also provide low-density paths for the passage of UV radiation from the disc to the halo. Other research by Davidge et al. (1998) examined the red giants in C72. They concluded that the oldest stars were around 10 billion years old and the distance to C72 was similar to other Sculptor cluster members, C70 (NGC300) and NGC7793 (i.e. about 7 million light years).

C73

Globular Cluster in Columba

Image

Buil-Thouvenot

Database

Name and/or Catalogue Designations:
NGC 1851
Dunlop 508
Bennett 32

Type of object: Globular Cluster

Catalogue position for epoch J2000.0
Right ascension: 05h 14m 06.3s
Declination: −40° 02' 50"

Constellation: Columba

Object information:
Magnitude: 7.3
Size: 11.0'
Object classification: 2
NGC Description: remarkable, very bright, very large, round, very suddenly very very
   bright middle, well resolved
Note: X-ray source.
Distance from the Sun: 12.2 kpc.
Distance from the galactic centre: 16.8 kpc.

Star atlas chart numbers:
Millennium Star Atlas, Charts 417–418, Volume I
Sky Atlas 2000.0, Chart 19
Uranometria 2000, Chart 393, Volume 2

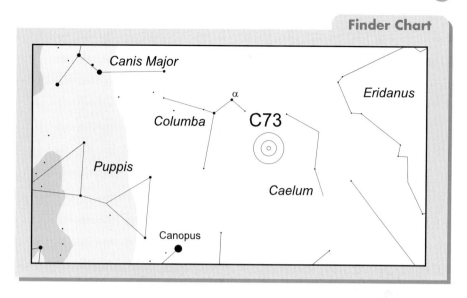

Visual Description

With C73 we make our only visit to the small constellation of Columba, where there is just one binocular deep-sky object, the globular cluster, NGC1851. In astronomical binoculars it looks like a 7th magnitude star with a hint of nebulosity. A 6 inch (150mm) telescope is sufficient to reveal its true identity although more aperture is required to resolve its brightest stars. These are between 13th and 14th magnitude so they are marginal even in an 8 inch telescope. What is apparent at this aperture is the very bright small central core giving it a 3-D appearance. Doubling the aperture then between 20 to 30 stars are resolved but not in the core where high magnification and steady seeing is essential (very unlikely if viewing the object low in the sky). It is located 8° south–west of Alpha Columbae (2.7 mag.) and visible in finderscopes.

Object Description

C73 is a globular cluster of considerable importance. There are 24 known RR Lyrae variable stars (Wehlau et al., 1978/82), it contains an X-ray burst source (Jernigan & Clark, 1979) and has an ultra-violet bright star, UV5 (Vidal & Freeman, 1975). If that was not enough, recent photometry (Saviane et al., 1994/5) has indicated that C73's morphology cannot be reproduced by "a single stellar population of fixed age". This is contrary to conventional theories, where globular clusters formed all their stars in one initial burst. Ultra-violet images of this cluster were taken with the Ultra-violet Imaging Telescope during the Astro-1 Mission on board the Space Shuttle (1990). Studies of these images by Parise et al. found that the star UV5 contributed up to 30% of the total UV flux for the cluster and was most likely a single star on the post-asymptotic giant branch (PAGB). They could find no optical counterpart for the X-ray source. From colour-magnitude diagram studies, Walker (1992) deduced the age of C73 to be around 14 billion years – uncomfortably close to the perceived age of the Universe!

C74

Planetary Nebula in Vela

Image

Steve Lee

Database

Name and/or Catalogue Designations:
The Eight-Burst Nebula
NGC 3132
PK272+12.1

Type of object: Planetary Nebula

Catalogue position for epoch J2000.0
Right ascension: 10h 07m 00.3s
Declination: −40° 26′ 40″

Constellation: Vela

Object information:
Magnitude: 8.2
Size: 1.4′ × 0.9′
Magnitude of central star: 10.1
Object classification: 4(2)
NGC Description: very remarkable planetary, very bright, very large, irregular extended, star 9th magnitude
Note: Southern Ring Nebula.
One of the brightest planetary nebula.

Star atlas chart numbers:
Millennium Star Atlas, Charts 941–942, Volume II
Sky Atlas 2000.0, Chart 20
Uranometria 2000, Chart 399, Volume 2

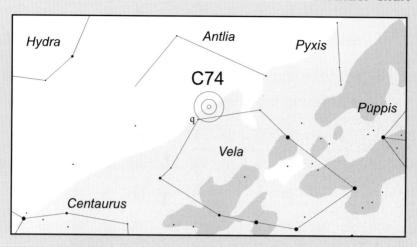

Nicknamed the Eight-Burst or Southern Ring Nebula, C74 does bear many similarities to M57, the Ring Nebula. Like M57, it is bright enough (estimates vary from magnitude 8.2 to 8.8) to be visible in an 8 inch (200 mm) telescope, even from suburban skies. It is more oval in shape than M57 and has perhaps a very slight blue tinge. A UHC filter improves the view showing a bright disc. It is however a bit smaller than its catalogue dimensions and I would suggest a size of 30 arc-seconds is nearer reality in small telescopes. What makes it unusual is its central star (see below), which is bright at magnitude 10 and therefore easily seen in a 6 inch (150 mm) telescope. Larger telescopes (12 inches–300 mm plus) show a distinct oval shape with the northern part of the nebula the faintest. The nebula is also visible in 20 × 80 binoculars appearing as a tiny faint glow of about 9th magnitude. C74 is located near to the border with Antila, in the northern part of Vela. The nearest guide star is q Velorum (mag. 4) – from this star aim 2.2° north-east. A Telrad finder with its 4° circle makes locating this object relatively easy.

Images taken with Hubble's WFPC2 have displayed this nebula in unprecedented detail. The central star was clearly shown to be binary but surprisingly it was the very faint companion (16th mag. at 1.6 arc-seconds separation), which is thought to be the one that is responsible for the nebula. This dwarf is now smaller than our own Sun but extremely hot, ionising the nebula and causing it to fluoresce. The more prominent central star is in a much earlier stage of stellar evolution and is simply not hot enough to be the culprit. Studies (Sahai et al., AAS) using the Hubble images have revealed new features including a host of filaments, one like a waistband across the ring, made of dust particles. These dust particles turned out to be rich in elements such as carbon. Who knows, one day these particles may be incorporated into new planets or even life. C74 is nearly half a light year in diameter and thought to be at a distance of 2000 light years.

C75

Image

Buil-Thouvenot

Database

Name and/or Catalogue Designations:
NGC 6124
Lacaille I.8
Dunlop 514
Melotte 145
Collinder 301

Type of object: Open Cluster

Catalogue position for epoch J2000.0:
Right ascension: 16h 25m 36.0s
Declination: −40° 40' 00"

Constellation: Scorpius

Object information:
Magnitude: 5.8
Size: 29.0'
Number of stars: 100
Magnitude of brightest star: 8.7
Object classification: II 3 m
NGC Description: cluster, bright, large, pretty rich, little compressed in the middle, stars 9th to 11th mag.
Note: discovered by Lacaille

Star atlas chart numbers:
Millennium Star Atlas, Charts 1461–1462, Volume III
Sky Atlas 2000.0, Chart 22
Uranometria 2000, Chart 407, Volume 2

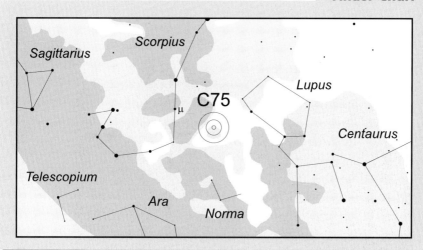

Visual Description

Near the Scorpion's tail are three open clusters, all different, but the one that concerns us here is NGC6124, C75. Discovered by Lacaille (alternatively M. l'Abbe de la Caille – see below) and published in his 1755 catalogue. His description was "it resembles a big comet without tail". His telescope was poorer than modern binoculars in which it appears as a soft hazy patch about half a degree across, not much dimmer than a 6th magnitude star. Its individual stars however start at about 9th magnitude and go fainter, so ordinary 50 mm binoculars struggle to reach just the brightest ones. Astronomical ones (20 × 80) however, show many member stars clearly, perhaps around 40, although it seems more. Stepping up to a 16 inch (400 mm) telescope then at least 100 stars are on view but all telescopes see a beautiful cluster with chains of stars, several doubles and a few red stars as well. It is located 5.75° WSW from Mu1 and Mu2 Scorpii, the wide double.

Object Description

C75 is a typical galactic open cluster with stars from magnitude 8.7 and it is one of the more nearby, at about 1,500 light years away. As referred to above, it was one of the discoveries of Abbe Lacaille, from the Cape of Good Hope, during his 1751–1752 journey to South Africa. It was included in his 1755 catalogue of 10,000 southern stars where he listed 42 "nebulous stars", including C75. This was some 16 years before Messier was to publish his more famous catalogue. The size of the telescope(s) he used was described, as was the fashion in that era, by its focal length of 15 to 18 inches. Its lens was probably therefore less than 1 inch (25 mm) in diameter! Lacaille divided his "nebulous stars" into three classes: Class 1: "Nebulosity not accompanied by stars"; Class 2: "Nebulosity due to clusters"; Class 3: "Stars accompanied by nebulosity". C75 was included as Class 1 No. 8. For more information on Lacaille see the article by O. Gingerich, Sky and Telescope February 1960.

C76

Open Cluster in Scorpius

Image

Buil-Thouvenot

Database

Name and/or Catalogue Designations:
NGC 6231
Melotte 153
Collinder 315
Dunlop 499

Type of object: Open Cluster

Catalogue position for epoch J2000.0
Right ascension: 16h 54m 00.0s
Declination: −41° 48′ 00″

Constellation: Scorpius

Object information:
Magnitude: 2.6
Size: 15.0′
Magnitude of brightest star: 4.7
Object classification: I 3 p n
NGC Description: cluster, bright, pretty rich, stars 10 to 13th mag.
Note: the surrounding region sometimes referred to as the Table of Scorpius or the Baby Scorpion

Star atlas chart numbers:
Millennium Star Atlas, Charts 1459–1460, Volume III
Sky Atlas 2000.0, Chart 22
Uranometria 2000, Chart 407, Volume 2

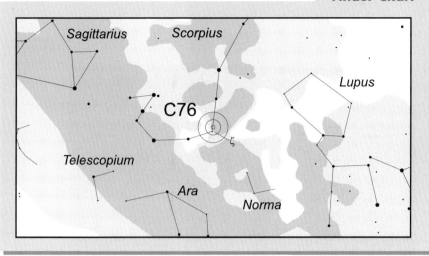

Visual Description

C76 is a fabulous sparkling open cluster in a fabulous part of the sky! Although its discovery is generally credited to Lacaille, it was presumably known since ancient times as it is a hazy naked-eye object. Its brightest stars are between 5th and 6th magnitude, which gives a better indication of its visibility rather than its misleading total magnitude of 2.6. A pair of binoculars is all that is needed to appreciate its majesty, showing 10 to 20 stars in a star-studded field. It has several similarities to the Pleiades, it is a young cluster and its prominent stars are blue-white, although not quite as bright it is still dazzling in a small telescope. Roughly a quarter of a degree in extent (but not round), a 6 inch (150 mm) telescope shows about 25 stars and averted vision shows the background haze of many many more. Its location is easy – in the tail of the Scorpion, just over half a degree north of Zeta Scorpii (mag. 3.6).

Object Description

The visual similarity of C76 to the Pleiades was referred to above but there are some significant differences. Its brightest stars are very high luminosity O and B supergiants, making it one of the brightest young clusters in the southern hemisphere. Generally considered to be part of the large stellar association, known as Sco OB1 (Bok et al. 1966), that encompasses several adjacent clusters as well. However, in 1984 Heske et al. suggested that C76 was not part of this association and was a foreground object, but in 1991, Perry et al., found no support for this. More recent research by Sung et al. (1998) has concentrated on photometry and they identified 12 pre-main-sequence stars and 7 possibles in the cluster. They also concluded that the massive stars seemed to have formed over a very short timescale during the latter stages of cluster formation. Research by Massa (1997, using HST images) has found the main sequence B stars to have extraordinary high stellar winds for which no satisfactory theory has yet been devised. C76's age is now generally accepted to be around 3 million years.

C77

Image

Buil-Thouvenot

Database

Name and/or Catalogue Designations:
Centaurus A
NGC 5128
ESO 270-9
ARP 153
PGC 46957

Type of object: Galaxy

Catalogue position for epoch J2000.0
Right ascension: 13h 25m 29.0s
Declination: −43° 01′ 00″

Constellation: Centaurus

Object information:
Magnitude: 6.8
Size: 18.2′ × 14.5′
Position angle: 35°
Object classification: S0pec
NGC Description: very remarkable, very bright, very large, very much extended P.A. 122°,
   bifid
Note: radio and X-ray source

Star atlas chart numbers:
Millennium Star Atlas, Charts 933–934, Volume II
Sky Atlas 2000.0, Chart 21
Uranometria 2000, Chart 403, Volume 2

Finder Chart

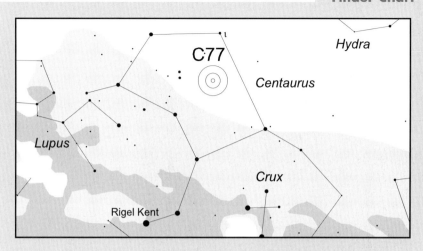

Visual Description

At magnitude 7 this is one galaxy that does not disappoint, even in a small telescope. To find it, move about 4° due north from the giant globular cluster Omega Centauri (C80), and then even a 6 inch (150 mm) telescope should reveal a round patch cut by a dark lane. There are reports of an 80 mm refractor being able to see this feature, although with averted vision. In a big Dobsonian (16 inch and upwards) the dark band takes on a mottled appearance. C77 is bright enough to be seen with binoculars, even when viewing it from the southern states of the USA, where it is low in the sky. It lies 6.3° south, and 1° east, of Iota Centauri (mag. 2.75).

Object Description

This enigmatic peculiar galaxy has puzzled astronomers for decades. At a distance of about 10 to 15 million light years this is the nearest active galaxy to us. What we see is believed to have resulted from the merger of a large elliptical galaxy and a smaller dusty spiral galaxy, about 1 billion years ago. The spiral galaxy deposited its gas and dust into the elliptical galaxy precipitating a flurry of star formation, beautifully revealed in a recent Hubble Space Telescope colour image. It is an immense emitter of radio waves (its name derives from being the brightest radio source in Centaurus) and at these wavelengths it has a bright jet, and fainter counterjet, both emanating form a compact core. Tingay et al. (1998) used Very Long Baseline Interferometry (VBLI) to examine the jets in radio waves and found relativistic speeds for them of around 10% the speed of light. This they took as adding evidence to the case for a massive black hole at the centre of C77. An accretion disc around this candidate was found to be 130 light years in diameter. However, probably the most surprising recent discovery is that by Mirabel et al. (1999) who found, right at the centre of the elliptical galaxy centred on the active nucleus, evidence for a barred type structure remarkably similar to that of a barred spiral galaxy.

Globular Cluster in Corona Australis

Alex Richter

Name and/or Catalogue Designations:
NGC 6541
Dunlop 473
Bennett 104
ESO280–SC004

Type of object: Globular Cluster

Catalogue position for epoch J2000.0
Right ascension: 18h 08m 02.2s
Declination: −43° 42′ 20″

Constellation: Corona Australis

Object information
Magnitude: 6.
Size: 13.1′
Object classification: 3
NGC Description: bright, round, extremely concentrated, gradually brighter towards the middle, well resolved
Note: stars from mag. 13.
Distance from the Sun: 7.4 kpc.
Distance from the galactic centre: 2.1 kpc.

Star atlas chart numbers:
Millennium Star Atlas, Charts 1457–1458, Volume III
Sky Atlas 2000.0, Chart 22
Uranometria 2000, Chart 409, Volume 2

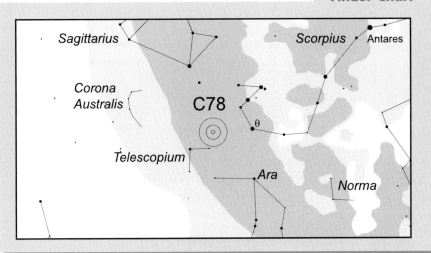

Visual Description

With C78 we return to another fine globular cluster which, due to its proximity to the "sting in the tail" of Scorpius, is located in a fine area for binocular sweeping. It is however "over the border" and in the constellation Corona Australis, which is not as conspicuous as its northern equivalent. It is located SE of a 5th magnitude star (magnitude revised by Hipparcos) and is an easy object in binoculars as it has a very bright core and a total magnitude around 6.5. That bright core is even more noticeable in a telescope, where an 12 inch (300 mm) is just able to resolve a few of the brighter members towards the outer reaches. These are around 13th magnitude so are not easy and require careful scrutiny. Its diameter is about 5 arc-minutes in this size of telescope. Member stars are more typically around 15th and 16th magnitude so a considerably bigger telescope is required for resolution of significant numbers. C78 is best located by sweeping 5.5° east from Theta Scorpii (mag. 1.86). Don't confuse it with a fainter (9th magnitude) globular (NGC6496) lying 1.7° WSW of C78.

Object Description

This bright globular cluster is about 24,000 light years from the Sun but less than 7,000 light years from the centre of our galaxy (Harris, 1997). Reference to its bright core was made above and it is in fact classed as a "core-collapsed" globular. Mendez et al. have used Hubble Space Telescope (HST) images of C78 to determine if the increased resolution of this telescope, over ground-based ones, would be sufficient to enable proper motions of cluster members to be detected. Their measurements of the 500 brightest stars in C78 led them to conclude that, with images taken only 5 or 6 years apart, internal motion could be revealed using the HST. This would greatly increase our knowledge of the poorly understood internal kinematics of globular clusters.

C79

Image

Buil-Thouvenot

Database

Name and/or Catalogue Designations:
NGC 3201
Dunlop 445

Type of object: Globular Cluster

Catalogue position for epoch J2000.0
Right ascension: 10h 17m 36.8s
Declination: −46° 24′ 40″

Constellation: Vela

Object information
Magnitude: 6.8
Size: 18.2′
Object classification: 10
NGC Description: very large, irregular round, little compressed towards the middle, stars
    13th…16th magnitude
Note: a retrograde globular cluster.
Distance from the Sun: 5.1 kpc.
Distance from the galactic centre: 8.9 kpc.

Star atlas chart numbers:
Millennium Star Atlas, Charts 961–962, Volume II
Sky Atlas 2000.0, Chart 20
Uranometria 2000, Chart 399, Volume 2

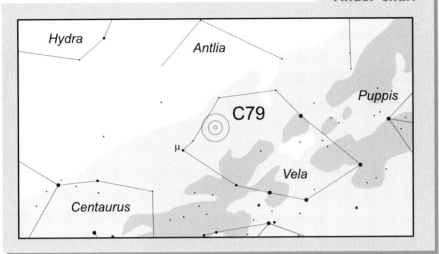

Visual Description

Discovered by Sir John Herschel at the Cape of Good Hope with an 18-inch f/13 speculum telescope, C79 is one of the more loose globular clusters but it is bright nevertheless. Just too faint for naked-eye visibility, it is an easy object in 50 mm binoculars as a hazy patch with a brighter middle. In astronomical binoculars it is quite large at around 8 arc-minutes but a telescope is needed to resolve its stars, which are from 12th magnitude and fainter. An 8 inch resolves many of them with averted vision but, to see it at its best, a 12 inch (300 mm) or larger is required where around 100 stars are visible across it face, without the concentrated core more typical of globulars. It is visible from the Southern USA but has been virtually ignored by northern observers. Located in the constellation Vela (the Sails), it is 5.75° north–west from Mu Velorum.

Object Description

C79 might just be regarded as yet another globular cluster but it has one peculiarity. It is going around our galaxy in the opposite direction of the general rotation of the galactic disc! In other words its orbit is retrograde. What event caused this is unknown. The internal dynamics of C79 have been investigated by Cote et al. (1995), who carried out an analysis of radial velocities for 399 giant stars in the cluster. They found some evidence for cluster rotation, which is supported by cluster ellipticity compatible with a rotationally flattened oblate spheroid, although alternative explanations are possible. On the other hand, Wallerstein et al. (1998) studied the elemental abundancies in C79. Their conclusion was that, despite its retrograde orbit, the composition of its stars was very similar to typical halo globulars. It is however rich in RR Lyrae variable stars.

C80

Globular Cluster in Centaurus

Image

Tim Puckett

Database

Name and/or Catalogue Designations:
Omega Centauri
NGC 5139
Lacaille I.5
Dunlop 440

Type of object: Globular Cluster

Catalogue position for epoch J2000.0
Right ascension: 13h 26m 45.9s
Declination: −47° 28′ 37″

Constellation: Centaurus

Object information
Magnitude: 3.7
Size: 36.3′
Object classification: 8
NGC Description: magnificent, extremely large, bright, extremely rich, very very compressed
Note: stars from magnitude 11 (Tycho Catalogue has one at 8.4 − foreground star?).
Distance from the Sun: 5.1 kpc.
Distance from the galactic centre: 6.3 kpc.

Star atlas chart numbers:
Millennium Star Atlas, Charts 953–954, Volume II
Sky Atlas 2000.0, Chart 21
Uranometria 2000, Chart 403, Volume 2

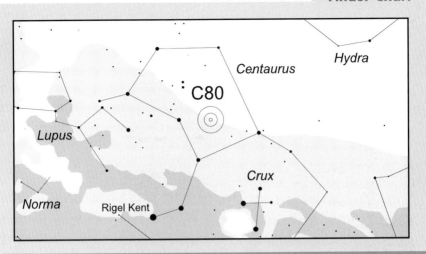

Visual Description

Simply the biggest and best globular cluster in the skies! Ptolemy plotted C80 in his Almagest nearly 2 millennia ago. Bayer catalogued it as a star (Omega Centauri) but it took the first telescope observer, no less than Edmund Halley in 1677, to realise it was a cluster of stars. With virtually any optical aid, that fuzzy naked eye "star" recorded by Ptolemy becomes a giant ball of stars. Realistically 20 × 80 binoculars are about the minimum needed to start resolving individual members. In a 6 or 8 inch (150–200 mm) telescope its size (at least that of the full Moon) and brightness are staggering. It is like observing any other globular with a 20 inch (500 mm) Dobsonian! Although the NGC lists it as "very very compressed", in reality this is not quite so and even a humble 8 inch telescope can resolve stars almost to the centre. Upping the aperture (12–16 inches), then the central area does become resolvable. Whatever the aperture, everywhere there is an underlying glow of literally millions of unresolved stars. Photographs can never capture its visual splendour of countless sparkling stellar pin-points.

Object Description

Omega Centauri is the brightest and most massive globular cluster in on our galaxy. It contains a large number of RR Lyrae stars, which has enabled its distance to be determined as 17,000 light years. The number of stars it comprises is not known precisely but is possibly up to 10 million. Meylan et al. determined the internal velocities of stars at various locations by measuring radial velocities over a 13 year period. The central velocity dispersion was the largest ever recorded and is indicative of a total mass of 5 million Suns. This is typically 10 times that of other globulars and is comparable to dwarf galaxies! The cluster is dominated by old red stars but the Ultraviolet Imaging Telescope (Nemiroff & Bonnell) revealed a substantial population of blue stars, stars which no longer fuse hydrogen to helium but are now fusing helium into carbon. Lynga (1996) was able to determine the exact centre of C80 (next to star 1587).

Globular Cluster in Ara

Image

Alex Richter

Database

Name and/or Catalogue Designations:
NGC 6352
Collinder 328
Melotte 170
Dunlop 417

Type of object: Globular Cluster

Catalogue position for epoch J2000.0
Right ascension: 17h 25m 24.0s
Declination: −48° 28′ 00″

Constellation: Ara

Object information
Magnitude: 8.1
Size: 7.0′
Magnitude of brightest star: 12.0
Object classification: 11
NGC Description: pretty faint, large
Note: listed in Sky Catalogue 2000 as both globular and open cluster. Saguaro Deep Sky
   database also lists it as an open cluster.
Distance from the Sun: 2.0 kpc.
Distance from the galactic centre: 8.64 kpc.

Star atlas chart numbers:
Millennium Star Atlas, Charts 1479–1480, Volume III
Sky Atlas 2000.0, Chart 22
Uranometria 2000, Chart 408, Volume 2

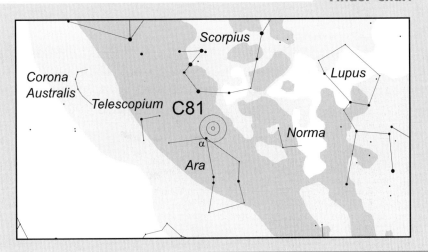

Visual Description

Moving further south into the constellation of Ara (the Altar) we come to another globular cluster C81 (NGC6352) and, if it were not following Omega Centauri (C80), it would be regarded as large and bright! But despite this it was missed by John Hershel but it is an impressive object in any telescope able to reach its 12th magnitude member stars. In an 8 inch (200 mm) telescope it appears about 4 arc-minutes across, not quite circular, with a slight brightening to the middle. The south side is slightly more prominent. It is also an easy object in large binoculars (20 × 80) where it stands out from neighbouring stars as a small hazy glow. It is located 1.75° north–west of Alpha Aræ (magnitude 2.8).

Object Description

Not a well known globular cluster but C81 does have one minor claim to fame! It is classed as one of the disc system of globulars (rather than the separate halo group) on the basis of its position, orbital motion and metallicity. That claim to fame? It was imaged by the Hubble Space Telescope *before* the installation of the corrective optics. These images were processed and analysed by Fullton et al. (1995). They were able to "deconvolve" the fuzzy images, i.e. restore them to what they should have been had the optics been perfect. To do this they used complex mathematical algorithms developed at the University of North Carolina. On the basis of their sharpened images, the age of C81 was put at slightly older than 47 Tucanae (C106) at around 14.5 billion years. However, in a more recent paper by Salaris et al. (1998) that age has been disputed. This team used up-to-date stellar models reducing C81's age (and C106) to 9.2 million years. Only time will tell which team is correct.

C82

Image

Buil-Thouvenot

Database

Name and/or Catalogue Designations:
NGC 6193
Collinder 310

Type of object: Open Cluster

Catalogue position for epoch J2000.0
Right ascension: 16h 41m 18.0s
Declination: −48° 46′ 00″

Constellation: Ara

Object information
Magnitude: 5.2
Size: 15.0′
Magnitude of brightest star: 5.7
Object classification: II 3 p n
NGC Description: cluster, very large, little rich, little compressed, involved in faint nebula
Note: in Ara OB1 and enveloped in the emission nebula NGC 6188

Star atlas chart numbers:
Millennium Star Atlas, Charts 1481–1482, Volume III
Sky Atlas 2000.0, Chart 22
Uranometria 2000, Chart 407, Volume 2

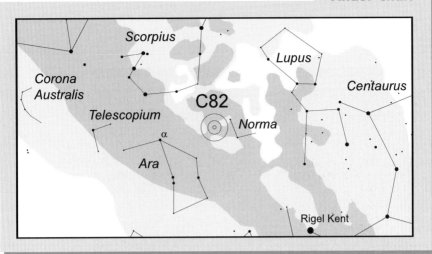

Visual Description

Two objects for the price of one! The open cluster C82 (NGC6193) and the associated emission/reflection nebula (NGC6188), sometimes known as the Rim Nebula. Marginally a naked-eye object, binoculars or modest telescope will provide an impressive view. It is a well spread out cluster, longer north–south than east–west. It has several interesting groupings and alignments of stars. There are about 30 to 40 stars around magnitude 9 and 10, and many more fainter ones. The brightest star, at the western side of the cluster, is an easy double (mag. 5 and 7 with 10 arc-seconds separation) and a tricky triple (separation 1.6 arc-seconds). This star is probably the main source of the illumination for the nebula. The latter, despite being prominent in photographs, is very illusive and, without a high contrast nebula filter, very difficult indeed. Although virtually on the galactic plane, there is no bright nearby star. Try heading 8.2° west from Alpha Aræ (magnitude 2.8) and then a touch north.

Object Description

C82 is part of the Ara OB1 star forming region and contains many young bright stars of types O and B, (hence the name) plus the remnants of the nebula from which they formed. It is located about 4,000 light years from the Sun. The complex around C82 is also known as RCW108 after Rodgers/Campbell/Whiteoak and their 1960 "Catalogue of H-alpha emission regions on the southern Milky Way". RCW108 is referred to as a molecular cloud and is in the process of being destroyed by intense ultra-violet light from the young hot stars in C82. This also causes the nebula to glow red, which makes it so prominent in David Malin's famous photograph of this region. Where the nebula is shielded from the radiation it is opaque and obscures the view. This is visible as a dark edge, hence the name Rim Nebula. Near the darkest part of the nebula is an infra-red source IRAS 16262-4845. Here, a cluster of new stars is currently being formed.

C83

Image

Steve Lee

Database

Name and/or Catalogue Designations:
NGC 4945
ESO 219–24
IRAS13025–4911
PGC 45279
Dunlop 411

Type of object: Galaxy

Catalogue position for epoch J2000.0
Right ascension: 13h 05m 26.2s
Declination: −49° 28′ 15″

Constellation: Centaurus

Object information
Magnitude: 8.6
Size: 20.0′ × 3.8′
Position angle: 43°
Object classification: SBc
NGC Description: bright, very large, very much extended in P.A. 39°
Note: part of Centaurus Galaxy Group NGC 4945, NGC 5068, NGC 5102, NGC 5128
& NGC 5236.

Star atlas chart numbers:
Millennium Star Atlas, Charts 953–954, Volume II
Sky Atlas 2000.0, Chart 21
Uranometria 2000, Chart 402, Volume 2

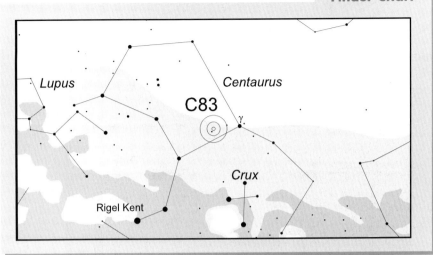

Visual Description

With C83 we leave the environs of the Milky Way and focus on a galaxy, which appears as a fainter version of C65 (NGC 253). It is a bit brighter than its magnitude implies but it can be very difficult in astronomical binoculars (20 × 80), where at best it is a subtle thin hazy sliver amongst a pretty, but distracting, foreground star field. An 8 inch (200 mm) telescope provides a much better view with the galaxy appearing about 10 arc-minutes long but still with not much detail other than perhaps a brighter core, slightly off centre. Its long axis is orientated north–east to south–west. Doubling the aperture almost doubles its visible length and shows a mottled surface, due to the large amount of dust in the galaxy. It is located 4° east from Gamma Centauri (mag. 2.2), with two stars (mag. 4 and 5) straddling it.

Object Description

C83, at first glance, is a rather unspectacular spiral galaxy seen about 6° from edge-on with its spiral structure difficult to discern. In part this is because the galaxy is itself extremely dusty and it is also seen through dust in the Milky Way. It is however now accepted to be a relatively nearby barred spiral type. But beneath that dusty surface much is going on. First it has a compact nucleus, which is extremely prominent at infrared and longer wavelengths. It also shows starburst features in the infra-red together with dense nuclear clouds, which give rise to both spectral-line absorption and emission features from a variety of different molecules. This includes the first detected H_2O "mega-maser" emission. It is one of the brightest Seyfert type 2 galaxies in X-rays (Done et al., 1996). Seyfert galaxies have extremely bright nuclei and are believed to be less powerful relatives of quasars. In Seyfert type 2 galaxies, the view of the central region is obscured possibly because of the presence of a thick molecular torus. In a collaborative project involving several radio astronomers over several years, Whiteoak et al. and Ott et al. (1996) confirmed the presence of the dense nuclear torus but, in addition, found evidence for a bar connecting the two spiral arms.

C84

Image

Alex Richter

Database

Name and/or Catalogue Designations:
NGC 5286
ESO220–SC038
Dunlop 388
Bennett 64

Type of object: Globular Cluster

Catalogue position for epoch J2000.0
Right ascension: 13h 46m 26.5s
Declination: −51° 22′ 24″

Constellation: Centaurus

Object information
Magnitude: 7.6
Size: 9.1′
Object classification: 5
NGC Description: very bright, pretty large, round, well resolved, stars 15th magnitude
Note: star M Centauri on SSE edge.
Distance from the Sun: 10.7 kpc.
Distance from the galactic centre: 8.2 kpc.

Star atlas chart numbers:
Millennium Star Atlas, Charts 971–972, Volume II
Sky Atlas 2000.0, Chart 25
Uranometria 2000, Chart 430, Volume 2

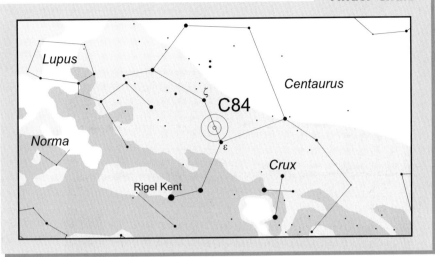

Visual Description

C84 is a bright globular cluster but everything is relative. Compared to nearby Omega Centauri (C80) it appears positively dim! It is visible in binoculars but with some difficulty. It is bright enough but the adjacent foreground star, M Centauri, makes it easy to locate but also easy to miss. C84 is NNW from M Centauri, which overlies the cluster's extreme edge. In a modest telescope (6 to 8 inches) the setting is delightful with unresolved glow of the globular contrasting with the sparkling yellow/orange of M Centauri. The individual stars of C84 are around 15th magnitude so are not seen at this aperture. Indeed they are only resolvable away from the centre and in considerably larger telescopes, although smaller apertures at highish power do reveal the uneven glow of clumps of stars. M Centauri is actually a double (it is also a spectroscopic binary – period 437 days) which is difficult due to the disparity in magnitudes of it two components (5 and 11) rather than its separation of 40 arc-seconds. It is located on a line joining Epsilon and Zeta Centauri, about 2.3° from Epsilon.

Object Description

Discovered by James Dunlop from Parramatta Observatory in 1827, C84 is a rather poorly studied globular cluster. In fact the first CCD derived colour-magnitude diagram (CMD) for it was not published until 1995 by Samus et al.. They carried out photo-electrically calibrated CCD BVRI photometry for 1,391 stars in the cluster. This enabled the resulting CMD to reach beyond the main sequence turn-off with sub-giants, horizontal branch and upper main sequence parts all well represented for the first time. Based on standard stellar models, an age of 15 to 17 billion years is indicated, a result that is difficult to reconcile with current estimates for the age of the Universe. In 1997, Garashchenko et al. also derived the CMD for C84, looking in particular for stars in the "instability strip". They were able to confirm the identification of 10 RR Lyrae type variables with the possibility of 8 more candidates.

C85

Image

Steve Lee

Database

Name and/or Catalogue Designations:
Omicron Velorum Cluster
IC 2391
Lacaille II.5
Collinder 191

Type of object: Open Cluster

Catalogue position for epoch J2000.0
Right ascension: 08h 40m 12.0s
Declination: −53° 04′ 00″

Constellation: Vela

Object information
Magnitude: 2.5
Size: 50.0′
Number of stars: 30
Magnitude of brightest star: 3.6
Object classification: II 3 p
NGC Description: very large, bright
Note: includes Omicron Velorum, a spectroscopic binary of period 3.2 hours, magnitude
 variable 3.51 to 3.56 and a period of 2.8 days (Hipparcos)

Star atlas chart numbers:
Millennium Star Atlas, Charts 981–982, Volume II
Sky Atlas 2000.0, Chart 25
Uranometria 2000, Chart 425, Volume 2

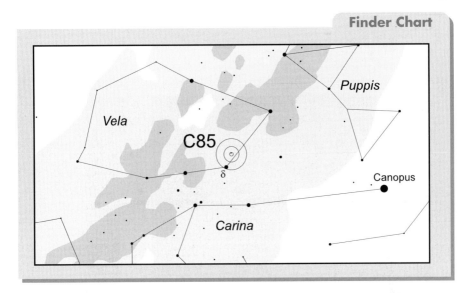

Visual Description

To the west of the False Cross lies the beautiful binocular object, the open cluster C85 (IC 2391). It is the brightest cluster in Vela and from a dark site it is just visible to the naked-eye as a couple of stars in a hazy patch. Any modest binoculars will show the brightest star, Omicron Velorum at magnitude 3.6, and up to about 30 other stars spread over quite a large area of 1°. Astronomical binoculars (20 × 80) are probably ideal as it is not a rich or concentrated cluster so any telescope (other than a richest field type) just spreads it out and in fact makes it less impressive. In those big binoculars the cluster contrasts magnificently with the smaller, unresolved adjacent cluster NGC 2669, less than one degree away on its eastern side. Some of the member stars are double and with sufficient separation for these binoculars to split them. A telescope surprisingly adds only a few extra stars to C85 as it doesn't appear to have many, if any, faint members. The cluster is to be found 1.75° NNW from Delta Velorum (mag. 1.93).

Object Description

The discovery of the open cluster is generally credited to Lacaille but it was first described by the old Persian astronomer, Al Sufi, in about 964 AD. It is a young cluster with an age of about 30 million years and recent Hipparcos measurements have put its brightest members at between 450 and 500 light years away. Mention above was made to the few faint members visible but Patten & Pavlovsky (1999) have searched C85 for brown dwarfs, objects just too small to sustain nuclear fusion. They identified 2 dozen cluster members, which were likely candidates as they matched the theoretical predictions of what 30 million year old brown dwarfs would look like. C85 has also been studied in X-rays (Simon & Patten, 1998, ROSAT). The main source is believed to be coronal X-rays from its young stars, which are just joining the main sequence.

C86

Image

Alex Richter

Database

Name and/or Catalogue Designations:
NGC 6397
Lacaille III.11

Type of object: Globular Cluster

Catalogue position for epoch J2000.0
Right ascension: 17h 40m 41.3s
Declination: −53° 40′ 25″

Constellation: Ara

Object information:
Magnitude: 5.7
Size: 25.7′
Magnitude of brightest star: 10.0
Object classification: 9
NGC Description: bright, very large, rich, stars from 10th magnitude
Note: second nearest globular.
Distance from the Sun: 2.2 kpc.
Distance from the galactic centre: 6.0 kpc.

Star atlas chart numbers:
Millennium Star Atlas, Charts 1495–1496, Volume III
Sky Atlas 2000.0, Chart 26
Uranometria 2000, Chart 434, Volume 2

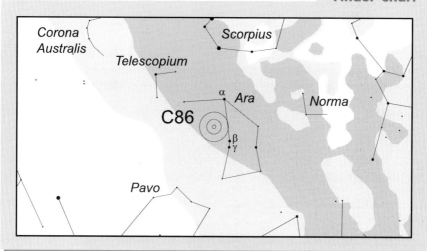

Visual Description

As one of the nearest globular clusters to us at approximately 7,200 light years (M4 is perhaps nearer at 6,800 l.y.), C86 is conspicuous even in binoculars. First spotted by Lacaille in 1755, it is relatively "open" for a globular cluster which, together with its proximity, makes resolution of its brighter stars fairly easy. It is also large, not much smaller than the full Moon. Even a modest 6 inch (150 mm) telescope will suffice to resolve its brightest stars as it has about twenty between 10th and 12th magnitude. These are randomly spread over the cluster, some appearing as chains, with no noticeable central condensation. The brighter stars are all red giants. Larger telescopes (12 inches and up) can resolve right to the core. The Hubble Space Telescope (HST) could even see straight through it – see below. It is located close to the plane of the Milky Way and about 2° to the east of the line of Alpha, Beta, Gamma Ara.

Object Description

C86 is one of the most studied globular clusters because of its proximity. Recent HST images, taken of just part of the cluster, were searching for red dwarfs but produced surprising results. No stars were found with a mass of less the $1/5^{th}$ the Sun. If there had been any then around 500 should have been visible in the Hubble field but none were visible. The stellar density was so low that background galaxies could be seen straight through it! Another recent study (Boesgaard/Thorburn 1997) has produced some strange results for lithium abundancy in the cluster – it is not uniform from star to star, yet globular cluster stars all formed together from the same material. Their explanation for this is that perhaps stars are born with differing spin rates – the faster ones mixing themselves up so that the amount of lithium detectable on their surfaces could be different.

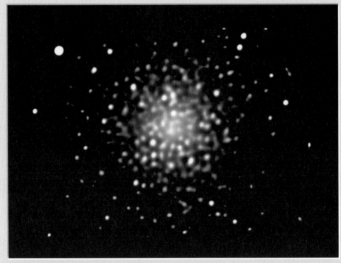

Buil-Thouvenot

Database

Name and/or Catalogue Designations:
NGC 1261
ESO155–SC011
Dunlop 337

Type of object: Globular Cluster

Catalogue position for epoch J2000.0
Right ascension: 03h 12m 15.3s
Declination: −55° 13' 01"

Constellation: Horologium

Object information
Magnitude: 8.4
Size: 6.9'
Object classification: 2
NGC Description: bright, large, round, partially resolvable
Note: magnitude 9.09 star outside north–east edge.
Distance from the Sun: 16.0 kpc.
Distance from the galactic centre: 17.9 kpc.

Star atlas chart numbers:
Millennium Star Atlas, Charts 459–460, Volume I
Sky Atlas 2000.0, Chart 24
Uranometria 2000, Chart 419, Volume 2

Finder Chart

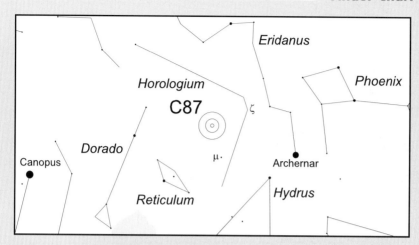

Visual Description

With C87 we make our only entry into the faint constellation of Horologium, the Clock, one of the constellations devised by Lacaille. C87 is in fact the brightest deep sky object in the constellation, being a reasonably bright globular cluster at magnitude 8.4. Astronomical binoculars (20 × 80) show it well together with a 9th magnitude star just outside its north–east edge. An 8 inch telescope (200mm) is insufficient aperture to resolve individual member stars but gives a pleasing view with its size somewhat smaller than its catalogue diameter. It appears about 3 arc-minutes across with a gradually brighter middle. Its individual stars start at magnitude 15 so a large aperture is required to resolve them, although there are reports (Hartung, who is notoriously optimistic!) that a 12 inch (300 mm) telescope is sufficient, even towards the centre. To locate C87 is a bit tricky as the guide stars Mu and Zeta Horologii are only magnitude 5.2. C87 is 4.5° from both and makes the right angle of a triangle with them, on their north-eastern side.

Object Description

C87 is a remarkably ordinary globular cluster but unlike C81, which is one of the disc group of clusters, C87 is a member of the Halo group. Recent studies of this cluster have been carried out by Ferraro et al. and Bolte et al., whose main findings were that it was very similar to the "standard" globular M5 both in metallicity and age. Perhaps more interestingly, C87 was one of the test objects for the European Southern Observatory's new SOFI instrumentation in 1998. The acronym stands for Son of Isaac and it is a state-of-the-art megapixel array for imaging in the infrared. It is the infra-red equivalent of a CCD. Used on their New Technology Telescope (NTT), it was able to record, in only 12 seconds, a stunning high resolution image centred on a wavelength of 1.65 microns. The resolution was 0.7 arc-seconds, excellent at this wavelength.

C88

Open Cluster in Circinus

Buil-Thouvenot

Database

Name and/or Catalogue Designations:
NGC 5823
Collinder 290
Melotte 131

Type of object: Open Cluster

Catalogue position for epoch J2000.0
Right ascension: 15h 05m 42.0s
Declination: −55° 36′ 00″

Constellation: Circinus

Object information
Magnitude: 7.9
Size: 10.0′
Number of stars: 100
Magnitude of brightest star: 9.7
Object classification: III 2 m
NGC Description: cluster, considerably large, rich, little compressed towards the middle,
    stars 13th to 14th magnitudes
Note: NGC 5822 lies 1.25° to the north

Star atlas chart numbers:
Millennium Star Atlas, Charts 967–968, Volume II
Sky Atlas 2000.0, Chart 25
Uranometria 2000, Chart 431, Volume 2

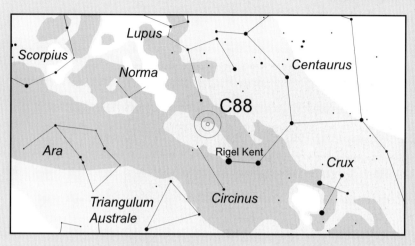

Visual Description

We enter another new constellation for us, Circinus (the Compasses), to track down the open cluster C88, NGC 5823. It is to be found right on the border with Lupus and is immediately south of the brighter and larger open cluster NGC 5822. If sweeping with binoculars it is NGC 5822 that will catch the eye, C88 is more subtle but it is there, about 1.25° south of its bigger companion. For northern observers I would compare it to M35 and its companion NGC 2158. C88 reveals its charms the larger the telescope becomes. Its true members number about 100 and are fairly faint at magnitude13 or fainter. It does however, have several overlying foreground stars (about 10 between magnitude 10 and 11) so even a small 6 inch (150 mm) telescope will resolve these. The background haze from the true cluster, about 8 arc-minutes in diameter, sets them off well. To find C88, sweep 7°NNE from Alpha Centauri (Rigel Kentaurus) and look for NGC5822 first.

Object Description

Not a well studied cluster, particularly not until recently. It was investigated by Bruck et al. (1968), who concluded that the cluster was probably associated with NGC 5822. Dawson (1978) cast doubt on the validity of C88 being a true cluster on the basis a study of red giants in the region. However, probably the best investigation of C88 was by Janes (1981). Whilst not answering all questions it did resolve several issues. His photometry revealed that C88 was a rather normal disc cluster, relatively young with possibly few red giants. Regarding the theory (Bruck) that it was associated with NGC 5822, the conclusion reached was that this was probably not the case and it was more likely that C88 was well beyond NGC 5822. Several of the foreground stars overlying C88 have had their distances measured by Hipparcos and were found to be in the range 40–200 light years. One 11th magnitude star did however have a distance similar to that postulated for the cluster of around 2,000 light years.

Open Cluster in Norma

Buil-Thouvenot

Database

Name and/or Catalogue Designations:
S Normae Cluster
NGC 6087 (see note below)
Melotte 141
Collinder 300

Type of object: Open Cluster

Catalogue position for epoch J2000.0
Right ascension: 16h 18m 54.0s
Declination: −57° 54′ 00″

Constellation: Norma

Object information
Magnitude: 5.4
Size: 12.0′
Number of stars: 40
Magnitude of brightest star: 7.9
Object classification: I 2 p
NGC Description: cluster, bright, large, little compressed, stars 7th to 10th magnitudes
Note: variable star S Normae at its centre.
C89 is NGC 6087 *not* NGC 6067 as originally listed

Star atlas chart numbers:
Millennium Star Atlas, Charts 1511–1512, Volume III
Sky Atlas 2000.0, Chart 26
Uranometria 2000 Chart 432, Volume 2

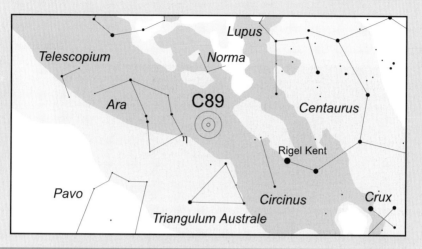

Visual Description

The first publication of the Caldwell Catalogue, and subsequent copies in various sources, unfortunately mis-identified the open cluster C89 as NGC 6067, rather than NGC 6087. The real C89 is also known as the S Normae Cluster after the famous Cepheid variable star (see below) in its midst. This is the brightest member of the cluster at magnitude 6.5 with the next brightest star at about magnitude 8. It is easily visible in a finder-scope or binoculars, as a loose irregular assemblage of about 10 stars. A modest telescope (8 inch–200 mm) shows about 40 stars down to 10th magnitude in an area about 12 arc-minutes in diameter. Those with a vivid imagination might make out patterns in the various lines of stars – arrow-heads and flowers have been seen! As Norma does not have any bright stars it is probably easiest to find C89 by heading 4.2° WNW from Eta Arae (mag. 3.8).

Object Description

Distance determination is a fundamental problem in Astronomy. The key to solving this problem is the use of "standard" candles of which Cepheid variables, such as S Normae, are the most important. Hipparcos has fixed the distance to the Pleiades and by comparing other open clusters with it, their distance can also be calculated. By doing this for C89 we will have determined the distance to S Normae, one of those standard Cepheid candles. Fry & Carney (University of NC, 1997) have worked on the photometry of C89 so that an accurate comparison to the Pleiades can be made, allowing for such factors as different stellar metallicity. Meanwhile Sagar & Cannon (1997) also used deep photometry of C89 to derive its age and distance. They concluded that S Normae was indeed a member, and, crucially therefore, at the same distance. That distance? Around 3,260 light years, with an age of 65 million years, is the best current estimate.

Planetary Nebula in Carina

Steve Lee

Name and/or Catalogue Designations:
NGC 2867
PK278–05.01
ESO126–PN008

Type of object: Planetary Nebula

Catalogue position for epoch J2000.0
Right ascension: 09h 21m 25.3s
Declination: –58° 18′ 43″

Constellation: Carina

Object information
Magnitude: 9.7
Size: 12.0″
Object classification: 4
NGC Description: very remarkable equal to 8th magnitude star, very small, round, star
    15th mag. near preceding (P.A. 90°), star near 13 arc-seconds
Note: magnitude of Wolf-Rayet central star is 14.9 visual and 16.6 photographic

Star atlas chart numbers:
Millennium Star Atlas, Charts 993–994, Volume II
Sky Atlas 2000.0, Chart 25
Uranometria 2000, Chart 425, Volume 2

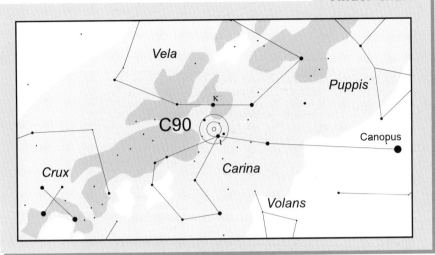

For those using binoculars, even astronomical ones, this object is quite a challenge. Not because it is too faint but because it is so tiny. Without adequate magnification (100x) it looks like an ordinary star. C90 was discovered by Sir John Herschel at the Cape of Good Hope, using an 18 inch (450 mm) telescope. He noticed no colour but most telescope users today agree it does have a slight blue tint. Despite being classed as a ring type planetary (Vorontsov-Velyaminov), this feature does not reveal itself in ordinary amateur telescopes, at least those up to 20 inches (500 mm). Its visual diameter is about 8 arc seconds – just about half the size of Mars at its biggest. The central star is too faint to see. The glow of the nebula reduces contrast so making its apparent brightness much fainter than its actual 15th magnitude might suggest. For those with equatorial telescopes it is easiest to locate by heading 3.3° due south from Kappa Velorum – otherwise head 1.1° NNE from Iota Carinae.

C90's discovery by Sir John Herschel has been referred to above but that is only half the story. Remember his father had discovered a planet, Uranus, 55 years previously. Not only that, this new planet was not keeping to its prescribed orbit and the intelligence of the day was that it was being perturbed by another one, as yet undiscovered. Imagine the excitement, when on the 1st April 1834, he discovered C90, which he described as "a perfect planet in appearance, with an attendant satellite". He fixed its position carefully against nearby stars. But understand his disappointment the following night when it had not moved. He had to conclude, "…is therefore not a planet". A suitable name therefore for C90 might be "Ghost of Neptune". Nevertheless, C90 is an interesting planetary nebula and is one of the few with a Wolf-Rayet central star. The latter is extremely hot (around 80,000K) and classed as type WC – the C standing for carbon, which is present in the form of spectral emission lines. An HST image is available (credit Bond) but with no associated caption.

Open Cluster in Carina

Image

Buil-Thouvenot

Database

Name and/or Catalogue Designations:
NGC 3532
Lacaille II.10
Melotte 103
Collinder 238
Dunlop 323

Type of object: Open Cluster

Catalogue position for epoch J2000.0
Right ascension: 11h 06m 24.0s
Declination: −58° 40′ 00″

Constellation: Carina

Object information
Magnitude: 3.0
Size: 55.0′
Number of stars: 400
Magnitude of brightest star: 6.07
Object classification: II 1 m
NGC Description: very remarkable, cluster, extremely large, round, little compressed, stars 8th to 12th magnitudes
Note: about 200 stars to magnitude 13

Star atlas chart numbers:
Millennium Star Atlas, Charts 991–992, Volume II
Sky Atlas 2000.0, Chart 25
Uranometria 2000, Chart 427, Volume 2

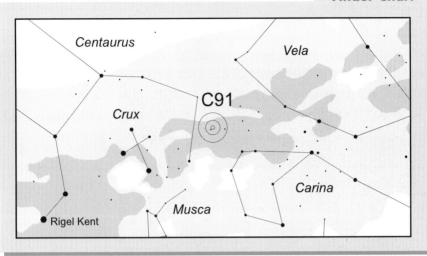

Although C91 was not one of Sir John Herschel's discoveries (Lacaille again!) it was one of his favourites, describing it as one of the most brilliant he had ever seen. One quick look with any optical aid will show why! Even small binoculars display a sparkling open cluster in the midst of the Milky Way, whilst larger ones (20 × 80s) will reach as many as 50 stars in an area about 45 arc-minutes E-W and 25 arc-minutes N-S. A feature of the cluster is that as optical aid gets larger the number of stars just keeps increasing, with around 150 in the range of even a modest telescope, such as an 8 inch (200 mm). Just outside the cluster, on the south-east is the orange tinted star, V382 Carinae at magnitude 3.9. All the visible cluster members (with one exception) are brilliant white. That one exception is its brightest star, the 6th magnitude HD96544, which also has an orange tint and is towards the north-east of the cluster. Whichever telescope is used, the widest possible field is essential. Poor Herschel, viewing long before the age of wide-field eyepieces, probably never saw the whole cluster at once. C91 is situated in the rich southern Milky Way and is clearly visible to the naked-eye as a distinct patch 2.5° east of the brilliant Eta Carinae Nebula (C92), 10° west of the Southern Cross.

C91 is a very young cluster, clearly older than the Pleiades but much younger than the Hyades. Its actual age is thought to be somewhat younger than 10 million years old. Around 400 stars are known to be true members with the cluster unusually rich in bright A-type stars. Shapley in 1930 reported 189 of the brighter stars were class A, 7 were G-type stars and 8 were K-type stars, but none were of type M. Coming to more up-to-date research (1995), the Ultra-violet Imaging Telescope (UIT) carried out observations of the cluster. The UIT observers were searching for ultraviolet counterparts to X-ray sources as well as for white dwarfs and other faint hot stars (see C93).

C92

Image

Steve Lee

Database

Name and/or Catalogue Designations:
Eta Carina(e) Nebula
NGC 3372
Gum 33
RCW 53
Dunlop 309

Type of object: Bright Nebula

Catalogue position for epoch J2000.0
Right ascension: 10h 43m 48.0s
Declination: −59° 52′ 00″

Constellation: Carina

Object information
Magnitude: 6.2
Size: 2° × 2°
Object classification: emission nebula
NGC Description: remarkable great nebula Eta Carina
Note: Keyhole Nebula & nova-like variable star Eta Carinae involved, many dark lanes
    present

Star atlas chart numbers:
Millennium Star Atlas, Charts 991–992, Volume II
Sky Atlas 2000.0, Chart 25
Uranometria 2000, Chart 427, Volume 2

Finder Chart

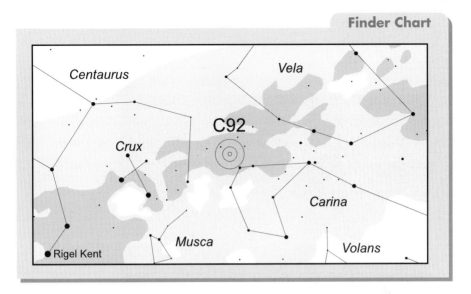

Visual Description

C92, the Eta Carinae Nebula, is perhaps the brightest of all parts of the Milky Way, easily visible to the naked-eye. Binoculars, such as 7×50 or 20×80, begin to show its huge extent (nearly 2°) and its complexity. With averted vision there is an obvious dark V-shape with the brighter part, containing Eta Carinae itself, to the north and each half having its own cluster of stars. Eta Carinae appears as an orange star with its own knot of nebulosity. It was observed by Sir John Herschel and he witnessed in 1838 its amazing brightening from 2nd magnitude to the 2nd brightest star in the sky. It could well perform again (there was an increase in April 1999) and it is always worth keeping an eye on it. Through a telescope only parts of the nebula can be examined at once but the nebula around Eta Carinae becomes more obvious and the whole field of view is filled with wisps of nebulosity. Herschel plotted 1200 stars in his drawing of the nebula, which gives some idea of his patience! C92 is around 13° west of the Southern Cross.

Object Description

C92, the Eta Carina(e) Nebula, is a giant H-II region lying in the Carina spiral arm of the Milky Way, at a distance of about 8,000 light years. The nebula glows from the excitation by many hot young stars of which Eta Carinae itself is the brightest and most enigmatic. It is an unstable giant around 100 times the mass of the Sun. According to Boumis et al. (1997) it is in its luminous blue variable stage, in which extreme mass loss is occurring in periodic violent outbursts. Its next phase could be to a Wolf-Rayet star before becoming a supernova. The nebula around the star, which masks its direct view, has been imaged by Hubble (1995). Revealed were two polar lobes and an equatorial disc. Smith & Gehrz plotted the lobes' expansion rate and confirmed their ejection began 160 years ago. However, Lamers et al. (1998) have produced evidence that Eta Carina could well be binary and have an unseen companion with a 5.5 year orbit. It was this companion, they claim, that ejected the nebula not Eta Carinae. Watch this space…

C93

Globular Cluster in Pavo

Image

Alex Richter

Database

Name and/or Catalogue Designations:
NGC 6752
Dunlop 295

Type of object: Globular Cluster

Catalogue position for epoch J2000.0
Right ascension: 19h 10m 51.8s
Declination: −59° 58' 55"

Constellation: Pavo

Object information
Magnitude: 5.4
Size: 20.4'
Object classification: 6
NGC Description: bright, very large, irregular round, stars 11th to 16th magnitude
Note: fine cluster.
Distance from the Sun: 3.9 kpc.
Distance from the galactic centre: 5.3 kpc.

Star atlas chart numbers:
Millennium Star Atlas, Charts 1507–1508, Volume III
Sky Atlas 2000.0, Chart 26
Uranometria 2000, Chart 435, Volume 2

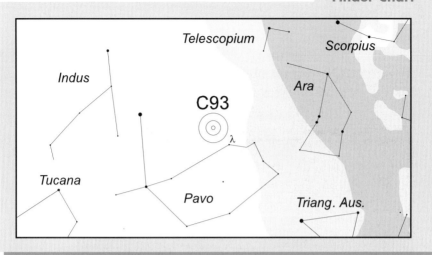

Visual Description

One of the top five southern hemisphere globular clusters! C93 is a beauty, just on the limit for naked-eye visibility in the constellation of Pavo. With optical aid, a bright 7th magnitude foreground star on the south-west of the cluster springs into view. Quite large and even 20 × 80 binoculars show it to be about 10 arc-minutes in diameter. In a telescope it does not appear perfectly circular leading to several observers (from Dunlop on) seeing it as two overlapping clusters (which is not the case). Adding to this illusion is the presence of some brighter stars as well as the main body of stars starting at about magnitude 13. It is one of the easiest clusters to resolve, with many of its stars appearing to flow in curving chains, leading to an unofficial name of the Starfish Cluster (ref. Marilyn Head). The core is very concentrated and even large (16 inch–400 mm) telescopes struggle to resolve it. It is situated 3.1° NE of Lambda Pavonis (mag. 4.2).

Object Description

C93 is a fairly typical globular cluster and contains over 100,000 stars. In 1995, it was imaged by Jim Sahli during the Astro-2 mission, on board the Space Shuttle Endeavour, using the Ultra-violet Imaging Telescope (UIT). Although the vast majority of C93's stars are not hot enough to emit detectable ultra-violet radiation, 355 stars were found to be so. These stars are believed to be in a short-lived stage near the end of their lives (known as the horizontal branch phase) when they can reach a surface temperature of up to 30,000 degrees. The UIT image was able to resolve the hot stars right into the cluster core. HST observations of C93 (Rubenstein & Bailyn 1997) have determined that the proportion of stars in the core of this globular cluster, which are binaries, is roughly 20–30%. Bailyn et al (1996), also using HST observations, identified 2 candidate cataclysmic variables near the core, one possibly coinciding with one of the X-ray sources (ROSAT) already known in the cluster.

C94

Open Cluster in Crux

Image

Steve Lee

Database

Name and/or Catalogue Designations:
Jewel Box or Kappa Crucis Cluster
NGC 4755
Melotte 114
Collinder 264
Lacaille II.12
Dunlop 301

Type of object: Open Cluster

Catalogue position for epoch J2000.0
Right ascension: 12h 53m 36.0s
Declination: −60° 20′ 00″

Constellation: Crux

Object information:
Magnitude: 4.2
Size: 10.0′
Number of stars: 200
Magnitude of brightest star: 5.8
Object classification: I 3 r
NGC Description: very remarkable cluster, very large, rich, very bright star (Kappa Crucis)
Note: 50 stars between magnitude 6 and 10 with many tinted ones.
Kappa Crucis (HIP62931) brightest member at mag. 5.9

Star atlas chart numbers:
Millennium Star Atlas, Charts 987–988, Volume II
Sky Atlas 2000.0, Chart 25
Uranometria 2000, Chart 429, Volume 2

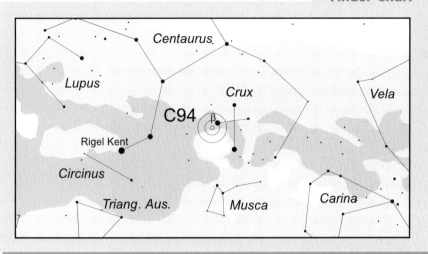

For most named deep sky objects you have to have a vivid imagination to see the resemblance to that name. That is not the case here. C94 more than lives up to its name, the Jewel Box. It resulted from Sir William Herschel's description of it as a "superb piece of fancy jewellery". It is just about visible to the naked-eye but it is in binoculars that its beauty first becomes apparent. They show a triangular shape of 6th magnitude blue-white stars but it is the prominent red one (HIP62918 – mag. 7.23) at its centre, which grabs the attention. Astronomical 20 × 80 binoculars give a superb view including the row across the middle of the triangle but being a very compressed cluster more magnification helps. Also the bigger the optical aid the more the individual colours become apparent. Herschel listed 8 stars with conspicuous colour (excluding the normal blue-white members) ranging from green to ruddy (red). The central row of stars is sometimes referred as the traffic lights because of their contrasting colours. C94 is located 1° south-east of Beta Crucis (mag. 1.25).

This spectacular young open cluster of around 200 members is dominated by one M-type (red) and four B-type (blue-white) supergiants. It has been much studied over the years but with considerable differences for its age and distance being postulated. What appears to make its distance and age difficult is differential reddening (obscuration) across the cluster. There seems to reasonable agreement now for its distance to be around 7,500 light years (Dachs/Kaiser, 1984 and Perry et al., 1976). This places it 300 light years above the galactic plane near the edge of the Carina-Sagittarius-Scutum spiral arm. Regarding its age, Sagar & Cannon (1995) determined this to be around 10 million years with most stars forming at the same time from a molecular cloud, which existed for about 6 to 7 million years.

C95

Open Cluster in Triangulum Australe

Image

Buil-Thouvenot

Database

Name and/or Catalogue Designations:
NGC 6025
Lacaille III.10
Melotte 139
Collinder 296
Dunlop 304

Type of object: Open Cluster

Catalogue position for epoch J2000.0
Right ascension: 16h 03m 42.0s
Declination: −60° 30′ 00″

Constellation: Triangulum Australe

Object information
Magnitude: 5.1
Size: 12.0′
Number of stars: 60
Magnitude of brightest star: 7.3
Object classification: II 2 p
NGC Description: cluster, bright, very large, pretty rich, irregular compressed, stars
    7th magnitude and fainter
Note: discovered by Lacaille in 1751–1752 from South Africa

Star atlas chart numbers:
Millennium Star Atlas, Charts 1511–1512, Volume III
Sky Atlas 2000.0, Chart 26
Uranometria 2000, Chart 432, Volume 2

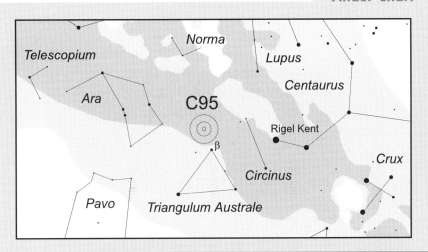

C95 is yet another discovery of the profligate Lacaille who, with his modest equipment (less than 1-inch–25 mm aperture), described it as "three faint stars in line in nebulosity". He was wrong about the nebulosity – there is none, but he was on the right track however with his "in line" description. Modest binoculars will show that it is not a round cluster and includes a linear alignment, mainly running north-east to south-west. However, it is at its best in a modest telescope. An 8 inch (200 mm) is about ideal as this can easily accommodate its quarter degree extent and is sufficient aperture to reach its faintest stars. There are about 50 from 8th to 12th magnitude visible and the irregular shape of the cluster quite obvious. A 6 inch (150 mm) will reach about 30 stars. It is just about visible to the naked-eye and the noted southern observer, Dunlop, referred to it as Lambda Circini. It is situated 3° NNE of Beta Trianguli Australis (mag. 2.83).

There have been various studies of this southern galactic cluster over the years with Feinstein (1971) publishing the first comprehensive review. He was able to differentiate between members and non-members by studying proper motion data and was then able to calculate its age on the basis of main sequence data, estimating it to be 100 million years. Kilambi (1975) added to this research using deeper photometry. His age was slightly younger at 90 million years and compared C95 to the Pleiades. The colour-magnitude diagrams for these two clusters proved to be very similar but C95 was found to have fractionally more bright stars than the Pleiades. We can only imagine how spectacular it would have been had it been at a similar distance away as the Pleiades (375 light years), instead of its 2200 to 2700 light years.

C96

Image

Buil-Thouvenot

Database

Name and/or Catalogue Designations:
NGC 2516
Lacaille II.3
Melotte 82
Collinder 172

Type of object: Open Cluster

Catalogue position for epoch J2000.0
Right ascension: 07h 58m 18.0s
Declination: −60° 52′ 00″

Constellation: Carina

Object information
Magnitude: 3.8
Size: 30.0′
Number of stars: 80
Magnitude of brightest star: 7.0
Object classification: I 3 r
NGC Description: cluster, very bright, pretty rich, stars magnitude 7th to 13th
Note: Burnham lists 3 doubles in cluster, h4027, h4031, i1104

Star atlas chart numbers:
Millennium Star Atlas, Charts 467–468, Volume I
Sky Atlas 2000.0, Chart 24
Uranometria 2000, Chart 424, Volume 2

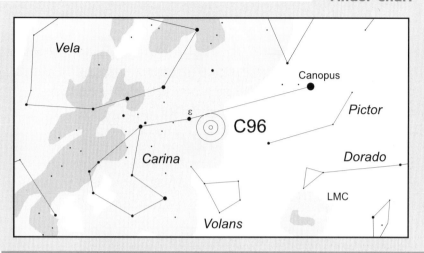

Visual Description

Sir John Herschel described C96 as "an orange-coloured star 8th magnitude, in middle of a large and magnificent cluster of perhaps 200 to 250 stars 8–16th magnitude many of the larger magnitudes, and really a superb object. Very visible to the naked eye". He was perhaps exaggerating the number of stars but he did have an 18 inch (450 mm) telescope. Such a big telescope is not needed to appreciate C96 and even 7×50 binoculars show about 20 stars scattered over a one degree field – Herschel was a bit pessimistic on star magnitudes here and even Lacaille saw 10–12 stars. In an 8 inch telescope, at low power, it is a superb sight with what looks like 100 stars and the red (orange?) giant at its heart unmissable. Also not to be missed are 3 double stars around magnitude 8 with separations 5 to 10 arc-seconds. Located just to the south of the False Cross and 3.2° WSW of Epsilon Carinae (mag. 1.86).

Object Description

C96 is not just a pretty face but also a very interesting young cluster with an age of about 150 million years. Jeffries et al. (1998) looked at the cluster's lithium content and stellar rotation. Several fast rotating, late G or early K type stars were seen. The pattern of lithium depletion was indistinguishable from that in the Pleiades, where the most rapid rotators suffer the least lithium depletion. Rotation periods for solar-type stars in NGC 2516 have been the subject of research by Barnes et al. using ROSAT X-ray images. They found evidence that most of the X-ray sources present emanate from young fast spinning solar-type stars, not the brightest stars. However, probably the most interesting new research (Zerbi et al., 1998) concerns the likely presence in C96 of 8 stars of the type known as Gamma Doradus. This new enigmatic variable class has only recently been recognised and about all that is agreed on for this class is that the phenomenon is a characteristic of relatively young F stars.

Image

Steve Lee

Database

Name and/or Catalogue Designations:
NGC 3766
Lacaille III.7
Melotte 107
Collinder 248
Dunlop 289

Type of object: Open Cluster

Catalogue position for epoch J2000.0
Right ascension: 11h 36m 06.0s
Declination: −61° 37′ 00″

Constellation: Centaurus

Object information
Magnitude: 5.3
Size: 12.0′
Number of stars: 100
Magnitude of brightest star: 7.2
Object classification: I 1 p
NGC Description: cluster, pretty large, pretty rich, pretty compressed, stars 8th to
    13th magnitude
Note: includes double star Inness 421 (7th and 10th mag., separation1.5 arc-seconds)

Star atlas chart numbers:
Millennium Star Atlas, Charts 989–990, Volume II
Sky Atlas 2000.0, Chart 25
Uranometria 2000, Chart 428, Volume 2

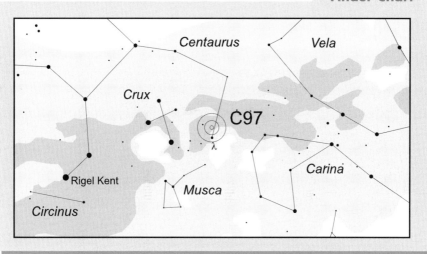

Finder Chart

Visual Description

The noted 19th Century southern observer Dunlop, described C97 as "a pretty large cluster of stars of mixed magnitudes, about 10' diameter. The greater number of the stars are of a pale white colour. There is a red star near the preceding side, another of the same size and colour near the following side, another small red star near the centre and a yellow star near the south following extremity....". It is a pretty accurate description of this fine open cluster set in a beautiful part of the Milky Way. With over 100 stars between 7th and 15th magnitude, good binoculars or a small telescope is all that is required to appreciate its charms. A pair of 20 × 80 binoculars reaches about a dozen stars and a 6 inch (150 mm) telescope over 50 with the two red stars, mentioned by Dunlop, standing out. It is one of those clusters where the eye joins stars up into patterns. It is to be found between Eta Carina and Crux, 1.4° north of Lambda Centauri (mag. 3.1).

Object Description

C97 is a young open cluster, located in the Carina complex, and Moitinho et al. (1997) using CCD photometry found it to be 25 million years old. However its claim to fame is that it contains an unusually large number of the mysterious Be type stars. Be stars are non-supergiant B-type stars whose spectra have, or had at some time, at least the H-alpha line in *emission*. The mystery of the "Be phenomenon" is that the emission, which is well understood to originate from a flattened circumstellar envelope or disc, can come and go on time scales of days to decades. Balona et al. (1991) studied the Be stars in C97 over a five year period. They found that, in spite of dramatic changes in the shapes and amplitudes of the light curves, the period was unchanged over that time. What is known about the evolution of these stars is that very young or old clusters have none or very few. They tend to peak in clusters in the age range of a few to 20 million years.

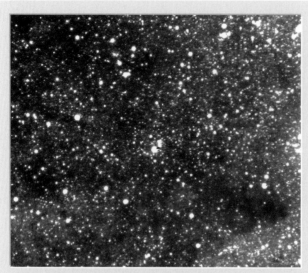

Buil-Thouvenot

Database

Name and/or Catalogue Designations:
Coalsack Cluster
NGC 4609
Collinder 263

Type of object: Open Cluster

Catalogue position for epoch J2000.0
Right ascension: 12h 42m 18.0s
Declination: −62° 58′ 00″

Constellation: Crux

Object information
Magnitude: 6.9
Size: 5.0′
Number of stars: 52
Magnitude of brightest star: 9.0
Object classification: II 1 p
NGC Description: cluster, pretty large, pretty compressed, considerably extended, stars
    10th magnitude...
Note: open cluster Hogg 15 is 12 arc-minutes to SE

Star atlas chart numbers:
Millennium Star Atlas, Charts 987–988, Volume II
Sky Atlas 2000.0, Chart 25
Uranometria 2000, Chart 451, Volume 2

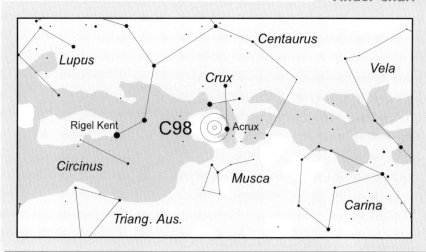

Visual Description

Not one of the most spectacular clusters but C98's location on the western edge of the Coalsack (C99) makes for a very interesting setting and of course has resulted in its common name of the Coalsack Cluster. It was discovered by Sir John Herschel with his 18 inch (450 mm) telescope and he recorded about 40 stars within it. However, with C98 having only 5 stars brighter than 11th magnitude, 20 × 80 binoculars will show it but not impressively. It just appears as a soft haze, with averted vision being needed to even tease out those 5 stars. An 8 inch (200 mm) telescope does rather better seeing about 20 stars but it is still rather sparse with no obvious concentration and somewhat elongated north-south. What is worth looking for is the nearby cluster Hogg 15, which is 12 arc-minutes to the south-east of C98, but with low power can be accommodated in the same field of view. This cluster is much smaller (2 arc-minutes in diameter) and with its brightest star magnitude 12, needs a 12 inch (300 mm) or larger to see it at its best. There is a 5th magnitude star between the two clusters, which makes the detection of Hogg 15 more difficult. C98 is located (obviously next to the Coalsack) and 1.8° due east of Acrux, Alpha Crucis (magnitude 0.77).

Object Description

C98 is a very little researched cluster with the definitive study as long ago as 1971 by Feinstein & Marraco, who placed the cluster beyond the Coalsack dark cloud. Coming more up-to-date, a study was carried out by Claria et al. (1998) of metallicity of high luminosity red stars in several young clusters, including C98, They examined the only such candidate star that Feinstein had identified in C98 but were able to confirm that it was probably a cluster member. C98 is about 5000 light years distant with an age of around 36 million years.

Dark Nebula in Crux

Steve Lee

Database

Name and/or Catalogue Designations:
Coalsack

Type of object: Dark Nebula

Catalogue position for epoch J2000.0
Right ascension: 12h 53m 00.0s
Declination: −63° 00′ 00″

Constellation: Crux

Object information
Magnitude: none
Size: 7° × 5°
Object classification: 3? Ir
NGC Description: none
Note: adjacent to the Southern Cross

Star atlas chart numbers:
Millennium Star Atlas, Charts 1001–1002, Volume II
Sky Atlas 2000.0, Chart 25
Uranometria 2000, Chart 451, Volume 2

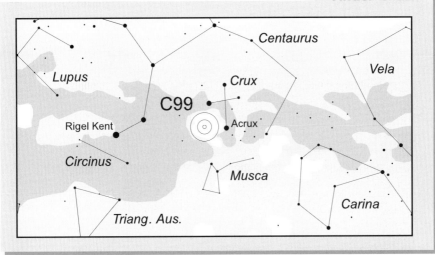

Visual Description

C99, the Coalsack, is our only dark nebula and, ignoring arguments over whether one can "see" a dark nebula or not, it is the most conspicuous in the skies and is easily visible to the naked-eye. It is perfect for wide field binoculars, such as 7 × 50s, which can accommodate its huge 7° × 5° size and looks for all as if a piece of the Milky Way is missing, a piece coincidentally about the size of the Large Magellanic Cloud. A telescope is really unsuitable but does demonstrate one important feature of it and that is that it is not devoid of stars at all. Any small telescope does in fact show more stars than many areas of sky well away from the galactic plane. The dark area is therefore an illusion and appears so dark because of the incredible brightness of the adjacent Milky Way. The brightest star in the middle of C99 is magnitude 6.6, which from a dark sky makes an excellent test of eyesight and sky transparency.

Object Description

Because of its position in the southern Milky Way and its nearness too us, the Coalsack dust cloud is the most prominent in the sky and has been the subject of several investigations since the 1930s when its distance was approximately determined (Unsold, 1929). The first reasonably accurate distance was determined by Rodgers (1960), who estimated it to be 570 light years away on the basis of the amount of stellar extinction observed in the stars visible within it. However, the most comprehensive study of it so far has been by Seidensticker & Schmidt-Kaler (1989). They produced a map of the cloud showing the varying extinction levels within it – it is far from uniform. Their explanation was to postulate at least *two* overlapping dark clouds (DCa & DCb) at distances of 600 and 800 light years with the possibility of a third at 475 light years. Perhaps the most unusual observation of the Coalsack was by the two Voyager probes (1994), which detected the ultra-violet scattering of starlight by the cloud.

C100

Image

Steve Lee

Database

Name and/or Catalogue Designations:
Lambda Centauri Cluster
(IC2944)
Collinder 249

Type of object: Cluster with Nebulosity

Catalogue position for epoch J2000.0
Right ascension: 11h 36m 36.0s
Declination: −63° 02′ 00″

Constellation: Centaurus

Object information
Magnitude: 4.5
Size: 60.0′ × 40.0′ (cluster & nebula)
Number of stars: 30
Magnitude of brightest star: 6.4
Object classification: II 1 p n
NGC Description: IC2944 "star 3.4 mag. in very large nebula"
Note: associated nebulae IC2944 & IC2948 (RCW62).
Inness 422 is a triple star in cluster of mags. 7.5, 10 and 12 (separations 0.4″, 1.7″, 9.6″)
    ref. Burnham.
Erroneously called Gamma Cen Cluster in original Caldwell list.

Star atlas chart numbers:
Millennium Star Atlas, Charts 1003–1004, Volume II
Sky Atlas 2000.0, Chart 25
Uranometria 2000, Chart 450, Volume 2

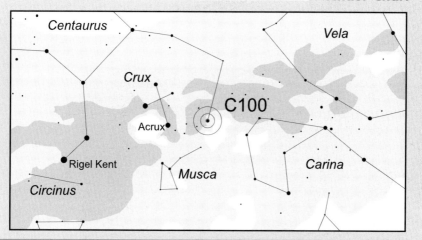

Every catalogue has its puzzling objects and C100 is certainly that. Firstly it was originally referred to as the Gamma Cen Cluster but the coordinates clearly refer to Lambda Cen. Secondly it is listed as IC 2944 but checking the original IC description clearly shows that this entry is referring to a nebula not a cluster. Enter Steve Lee, who checked the area both visually and on the UKST sky survey, and I am grateful for his detective skills. He found no cluster coincident with Lambda Cen but there was a sparse cluster 25 arc-minutes to the SE. With an O-III filter he could easily see that it was enveloped in extensive nebulosity. Checking the IC again he determined that this was IC2948. The nebula IC2944 around Lambda was too faint to see although it records well photographically. We checked through various papers and traced the first reference to the cluster SE of Lambda to Collinder who, in 1931, identified this aggregate of O stars (centred on HD101205) as Collinder 249. C100 should therefore be identified as Collinder 249. Binoculars will show it easily although a telescope and an O-III is required for the nebulosity (IC2948), which Steve could see even with a first quarter Moon present.

As referred to above, the cluster was first listed by Collinder (1931) and Alter et al. (1958) catalogued it as the Centaurus OB 2 association. It is embedded in an extensive H-II region (IC2944/8 – RCW62), which, ionised by the young O type stars, records spectacularly red in photographs. A feature of the nebula is the presence of several prominent Bok Globules displayed beautifully in a David Malin photograph but mis-labelled as IC2944 instead of IC2948. These globules, named after astronomer Bart Bok, comprise gas and dust clouds, which are clearly associated with the nebula (rather than in front) and are thought to be condensing into stars. The status of C100 as a true cluster was questioned by Perry et al. (1986) who produced photometric evidence to indicate that it was just a line of sight alignment. However, C100's status was restored later the same year when Walborn showed that the ionising O stars in C100 did constitute "a significant physical cluster". A more recent development concerns the puzzling "Thackeray's Globules", discovered nearly 50 years ago in IC 2944/8 by A.D. Thackeray. These small globules are located precisely on the line of the bright OB stars. Reipurth et al. concluded they were the remnants of a disintegrating "elephant trunk" in the nebula.

Galaxy in Pavo

Buil-Thouvenot

Name and/or Catalogue Designations:
NGC 6744
PGC 62836
ESO 104-42
IRAS19050-6354

Type of object: Galaxy

Catalogue position for epoch J2000.0
Right ascension: 19h 09m 45.4s
Declination: −63° 51′ 22″

Constellation: Pavo

Object information:
Magnitude: 8.3
Size: 15.0′ × 10.0′
Position angle: 15°
Object classification: SBb or SBc
NGC Description: considerably bright, considerably large, round, suddenly very much brighter to the middle, not well resolved
Note: several satellite galaxies including NGC6744A to the NW (15th mag.)

Star atlas chart numbers:
Millennium Star Atlas, Charts 1519–1520, Volume III
Sky Atlas 2000.0, Chart 26
Uranometria 2000, Chart 456, Volume 2

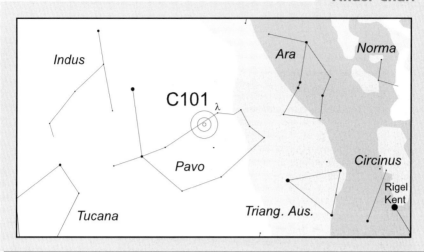

Visual Description

This is our last galaxy but one far too good to miss. C101 (NGC 6744) is probably the largest known barred spiral and, in a large telescope, one of the prettiest. It is inclined at about 50° to our line of sight so is quite open. Although listed at magnitude 9, binoculars struggle with it and even astronomical ones (20×80) only reach its brighter core with averted vision. Its large size (approaching 1/3rd degree in photographs) makes for a low surface brightness. Through a telescope it has 3 distinctive zones; a bright featureless inner core, about 2 arc-minutes long and aligned N-S, a middle zone of brighter spiral arms (about 10 arc-minutes) and an outer zone of very faint spiral arms. An 8 inch telescope can reach the first two and a 16 inch all three with the spiral arms, or more accurately the bright H-II and OB regions in them, being detectable. It is located 2.6° ESE of Lambda Pavonis (mag. 4.22) or alternatively head 4° due south from C93.

Object Description

C101's apparent size of 1/3rd degree taken with its distance of 34 million light years means that C101 is a giant. That size has counted against it making it too large to fit the field of view of most professional optical and radio imaging systems. However, a recent study by Ryder/Walsh/Malin (1999 in press), using both radio and optical images, has covered its extent by joining many images together in a mosaic. C101 is similar to our own galaxy but is relatively isolated having no major companions but they did find that one of its small satellites (NGC6744A) was interacting with it. Other findings were that the H-I (neutral hydrogen) gas extends out in two main spiral arms, 1.5 times the optical size, with the bulk of the material residing in a "ring" which underlies the outer optical disc. They also derived the rotation curve for C101, which is compatible with the galaxy having an unseen "dark matter" halo.

C102

Image

Buil-Thouvenot

Database

Name and/or Catalogue Designations:
Theta Carinae Cluster
IC 2602
Lacaille II.9
Melotte 102
Collinder 229

Type of object: Open Cluster

Catalogue position for epoch J2000.0
Right ascension: 10h 43m 12.0s
Declination: −64° 24′ 00″

Constellation: Carina

Object information
Magnitude: 1.9
Size: 50.0′
Number of stars: 60
Magnitude of brightest star: 2.8
Object classification: II 3 m
NGC Description: cluster, includes Theta Carinae
Note: at the edge of Sco/Cen Association

Star atlas chart numbers:
Millennium Star Atlas, Charts 1003–1004, Volume II
Sky Atlas 2000.0, Chart 25
Uranometria 2000, Chart 449, Volume 2

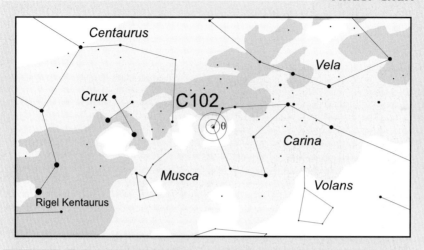

Visual Description

The open cluster C102 extends over 1° and consequently the best views of this southern gem are through either astronomical binoculars or a small (say 6 inch – 150 mm) richest field telescope (RFT). It is just 5° south of the galactic plane and is dominated by the bright star Theta Carinae (mag. 2.76). This is an early type star, nominally of spectral class BO, and is by some 2 magnitudes the brightest star in IC 2602. C102 contains over 80 members, 30 of which are brighter than 9th magnitude. Through that RFT the view is impressive with the cluster appearing in two parts, an eastern group with 5 prominent members (the "five of diamonds") and a western group around Theta. Larger telescopes, or higher magnifications, can spoil the view by spreading the stars out too much – so small is beautiful with C102! Theta Carinae is located 11° west of Crux.

Object Description

C102 is relatively young with an age around 30 million years. At a distance of only 500 light years away, it is one of the closest clusters, at the edge of the Scorpio-Centaurus Association. It has many similarities to the Pleiades but research by Randisch et al. (1995), using ROSAT X-ray data, found the cluster to be more X-ray luminous with 110 objects detected. They put this down to early stars in C102 rotating more rapidly than their counterparts in the Pleiades. In 1997, Patten et al. added evidence to this by deriving rotation periods for 9 K-type cluster members, which ranged from 0.20 to 4.5 days. Theta Carinae itself is a very interesting object(s) and is in fact a variable. Lloyd et al. (1995) studied its radial velocity variations using existing and new IUE (ultra-violet) data. They failed to confirm any of the orbital periods proposed for this system. Instead a new period of 2.2 days was found and they suggested that there is a second period at about 25 days. It could therefore be a binary or even a triple system.

C103

Image

Steve Lee

Database

Name and/or Catalogue Designations:
Tarantula Nebula
30 Doradus
NGC 2070
Lacaille I.2

Type of object: Bright Nebula (in LMC) with Cluster

Catalogue position for epoch J2000.0
Right ascension: 05h 38m 36.0s
Declination: −69° 05′ 00″

Constellation: Dorado

Object information
Magnitude: 8.3 (cluster only)
Size: 30′ × 20′ (photographic size 40′ × 25′)
Object classification: emission
NGC Description: magnificent, very bright, very large, looped
Note: located in Large Magellanic Cloud.
Part of LMC O-association No.100.

Star atlas chart numbers:
Millennium Star Atlas, Charts 495–496, Volume I
Sky Atlas 2000.0, Chart 24
Uranometria 2000, Chart 445, Volume 2

Finder Chart

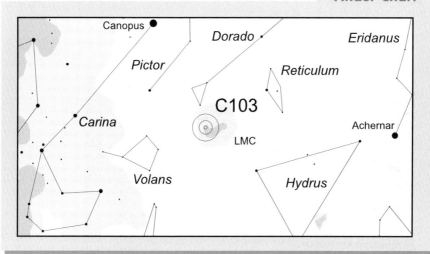

Visual Description

C103 is a nebula in another galaxy! It is amazing that it is just possible see a hint of it with the naked-eye, appearing as a brighter patch (it is brighter than its catalogued 8.3 magnitude) on the eastern side of the Large Magellanic Cloud (LMC). Observed extensively by Sir John Herschel he described it as an "assemblage of loops" and he catalogued 105 stars within it. Astronomical binoculars (20 × 80) show it as a slightly oval patch but it is quite small at this aperture. It is after all 180,000 light years away! It improves markedly in large telescopes and 10 inch (250 mm) and upwards begin to show those loops that Herschel referred to (the tarantula's legs). The 10th magnitude "star", 30 Doradus, is also evident and is a touch south of the middle. Surrounding it are many fainter ones (mag. 12–14). Some nebula filters improve the seeing of Herschel's loops and, as C103 has a strong Oxygen emission (501nm), an O-III filter should work best.

Object Description

An exceedingly complex object comprising a super-giant H-II region (the largest known anywhere!) with a dense central cluster formerly thought to be a single massive star, 30 Doradus. It has been extensively studied by several teams, using the resolution of the Hubble Space Telescope, to try and reveal the secrets of its central core. That core is known as Radcliffe 136 (R136) and from ground based images could only be resolved into 3 components with the largest, R136a thought to be an impossibly massive star equivalent to 2,500 Suns! It took Hubble to restore some sanity, showing R136a to be a super dense cluster containing thousands of massive hot blue stars. This cluster is just as puzzling as its age (about 3.5 million years) causing it to be described as an open cluster but its morphology is more akin to a globular cluster. Walborn et al. (1999) also report on several exotic objects inside 30 Doradus triggered by R136 including infra-red sources, jets, knots, dust pillars (like M16) and an H_2O maser.

C104

Image

Alex Richter

Database

Name and/or Catalogue Designations:
NGC 362
Dunlop 62

Type of object: Globular Cluster

Catalogue position for epoch J2000.0
Right ascension: 01h 03m 14.3s
Declination: −70° 50′ 54″

Constellation: Tucana

Object information
Magnitude: 6.6
Size: 12.9′
Object classification: 3
NGC Description: very bright, very large, very compressed, very much brighter towards the middle
Note: overlying the north of the Small Magellanic Cloud.
Distance from the Sun: 8.3 kpc.
Distance from the galactic centre: 9.2 kpc.

Star atlas chart numbers:
Millennium Star Atlas, Charts 501–502, Volume I
Sky Atlas 2000.0, Chart 24
Uranometria 2000, Chart 441, Volume 2

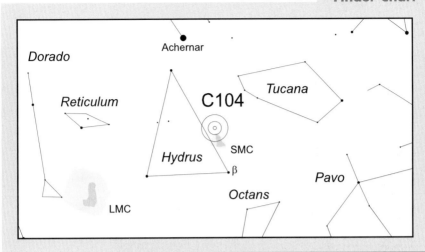

Visual Description

At the northern end of the Small Magellanic Cloud (SMC), we find a bright globular cluster, C104 (NGC 362). Appearances can be deceptive, this cluster is actually in our Milky Way galaxy and only coincidentally appears close to the SMC. This 4 arc-minute globe is easily visible in binoculars and, if you can hold them steady enough, shows a bright stellar core. All telescopes will show this but to resolve its 13th magnitude and fainter member stars, a 10 to 12 inch (250–300 mm) telescope is required. At this aperture the cluster's diameter grows a touch to 5 or 6 arc-minutes but a feature is the (apparent) lines of stars, almost to the central bright core. If the SMC is invisible, then to locate C104 head 7° NNE from Beta Hydri (mag. 2.82).

Object Description

C104 is a very famous, or probably more accurately infamous, globular cluster. It is one half of the classic "Second Parameter" puzzle duo. Paired with NGC 288 (another halo cluster), they have different "horizontal branches" on their HR diagrams, yet have the same metal abundancy (the first parameter). The horizontal branch differences in these two (C104 is redder) indicates that at least one other parameter must be causing this. But what is it! Some investigators have claimed significant age differences between the two clusters while others claim no age difference. Spectroscopic results have also proven disparate. Some have questioned the identical metallicity but recently M.D. Shetrone (McDonald Observatory) found both to be –1.3. He also produced evidence that other possibilities: deep mixing, oxygen depletion and sodium and aluminium enhancements, were the same in both clusters but Briley et al. (1994) have produced evidence that the stars in NGC362 have undergone more extensive deep mixing than their NGC 288 counterparts. The debate rages on!

C105

Image

Buil-Thouvenot

Database

Name and/or Catalogue Designations:
NGC 4833
Lacaille I.4
Dunlop 164

Type of object: Globular Cluster

Catalogue position for epoch J2000.0
Right ascension: 12h 59m 35.0s
Declination: −70° 52′ 29″

Constellation: Musca

Object information
Magnitude: 7.4
Size: 13.5′
Object classification: 8
NGC Description: bright, large, round, gradually?, very small bright middle, stars 12th mag.
Note: 3° north-east of C108 (NGC4372).
Distance from the Sun: 5.9 kpc.
Distance from the galactic centre: 6.9 kpc.

Star atlas chart numbers:
Millennium Star Atlas, Charts 1011–1012, Volume II
Sky Atlas 2000.0, Chart 25
Uranometria 2000, Chart 451, Volume 2

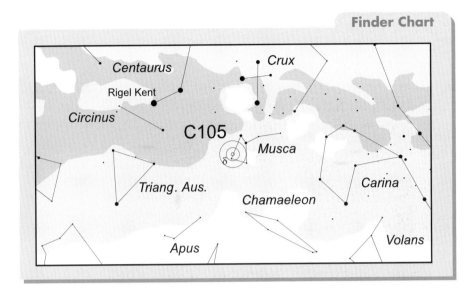

Yet another of Lacaille's discoveries. Using his small telescope, this globular cluster appeared to him as "a small faint comet". C105's morphological differences to C104 are evident in its visual appearance. It is a much looser, open aggregation, which results in a lower total magnitude yet paradoxically its individual stars are brighter. These start at around magnitude 12. It is nevertheless easy to track down with binoculars especially as it has a brightish (foreground) star, magnitude 8.7, overlying it just north of centre. An 8 inch (200 mm) is about the minimum to resolve its brighter stars but telescopes in the range 12 to 16 inches (300–400 mm) in aperture enable the cluster to be resolved right to its centre. It is to be found 1° NNW of the binocular double Delta Muscae (magnitude 3.6). This star stands out in binoculars as an obvious double with the primary a touch yellow.

C105 has been classed over the years as a moderately metal-poor or a very low metal abundancy cluster! The latest results (Geisler et al., 1992) have come down on the very low metal abundancy scenario. To be fair, it is not an easy globular to study being only 8° from the plane of the Milky Way and therefore inevitably suffers from obscuration plus considerable contamination by field stars. There are a few early studies of C105 including one by Demers & Wehlau, who determined periods for nine variables and extended the number of the RR Lyrae type to 15 in C105. Most recently, Samus et al. (1995) produced a new deep colour-magnitude diagram (CMD) for C105 using images taken at the Cerro Tolodo, Las Campanas and European Southern observatories. The resultant CMD reached the main sequence turn-off point and produced indications of a red-giant branch "bump" not dissimilar to other low metallicity globular clusters.

C106

Image

Steve Lee

Database

Name and/or Catalogue Designations:
47 Tucanae
NGC 104
Lacaille I.1
Dunlop 18

Type of object: Globular Cluster

Catalogue position for epoch J2000.0
Right ascension: 00h 24m 05.2s
Declination: −72° 04′ 51″

Constellation: Tucana

Object information
Magnitude: 4.0
Size: 31′
Object classification: 3
NGC Description: very remarkable, very bright, very large, extremely rich, very much
    compressed towards the middle
Note: stars from magnitude 11.7.
Distance from the Sun: 4.3 kpc.
Distance from the galactic centre: 7.3 kpc.

Star atlas chart numbers:
Millennium Star Atlas, Charts 501–502, Volume I
Sky Atlas 2000.0, Chart 24
Uranometria 2000, Chart 440, Volume 2

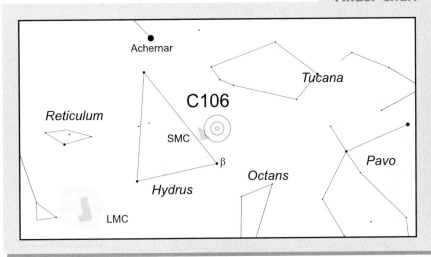

Visual Description

Second only to Omega Centauri, C106 (47 Tucanae) is a glorious 4th magnitude globular cluster and is one of the few that has a hint of colour. It is slightly smaller than Omega Centauri but differs in that it has a much more concentrated core. This core is even visible through 20×80 binoculars where it appears as a 1 to 2 arc-minute brighter central zone inside a fainter halo, about 10 to 15 arc-minutes in diameter. It gets better in a telescope where the core has a definite yellow tinge in contrast to the white outer area. This was first noticed by Sir John Herschel, who recorded the centre as ruddy or orange-yellow. As the telescope gets bigger the view gets even better with C106 almost 25 arc-minutes across with many of its 12th magnitude and fainter stars being resolved. There are a few foreground stars – look for a red one in the south. C106 is located just to the west of the Small Magellanic Cloud.

Object Description

This impressive globular cluster is one of the closest (about 15,000 light years), one of the heaviest (about 1 million solar masses) and has around I million stars. Mention was made above of its concentrated core and this seems to be just prior to a phase known as "core collapse". It has 5 X-ray sources, 11 millisecond pulsars, many eclipsing binaries and blue stragglers. One of the latter, known as BSS 19, was observed (Shara et al., 1997) using the HST and is the first blue straggler to have its mass determined (1.7 solar mass). Taken with its rapid spin this lends support for the theory that these stars form by the coalescence of a contact binary. Another strange star in C106 is AKO 9, which Meylan et al. (1997) witnessed to undergo a huge increase in brightness of 2.1 magnitudes in under one hour, briefly becoming the brightest star in the cluster. Whilst it is certain to involve a binary system the actual mechanism that resulted in AKO 9's behaviour is yet to be determined.

C107

Globular Cluster in Apus

Image

Alex Richter

Database

Name and/or Catalogue Designations:
NGC 6101
Dunlop 68

Type of object: Globular Cluster

Catalogue position for epoch J2000.0
Right ascension: 16h 25m 48.6s
Declination: −72° 12′ 06″

Constellation: Apus

Object information
Magnitude: 9.3
Size: 10.7′
Object classification: 10
NGC Description: pretty faint, large, irregular round, very gradually brighter to the middle, stars 14th magnitude
Note: a low central concentration cluster.
Distance from the Sun: 15.1 kpc.
Distance from the galactic centre: 11.0 kpc.

Star atlas chart numbers:
Millennium Star Atlas, Charts 1533–1534, Volume III
Sky Atlas 2000.0, Chart 26
Uranometria 2000, Chart 454, Volume 2

Finder Chart

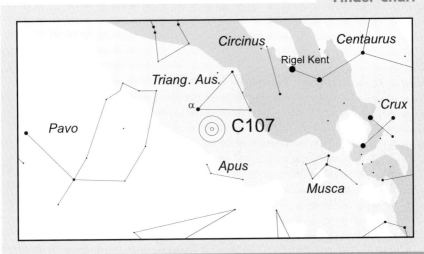

Visual Description

Not one of the brighter globular clusters, but C107 (NGC6101) is worth searching out particularly with large telescopes. It is on the limit of medium binoculars where it just appears as a 9th magnitude star unless they are astronomical ones with reasonable aperture and magnification e.g. 20 × 80s. These distinguish it from the surrounding stars, showing it as a faint hazy patch. Even small telescopes, 6 to 8 inch (150–200 mm), do not show much more, other than making its 3 to 4 arc-minute size more definite with a very slightly brighter centre. Its brightest stars are only a touch brighter than 14th magnitude so resolution at this aperture is optimistic. It really is an object for 12 inch (300 mm) or larger telescopes, where the light gathering power is sufficient to resolve its member stars. The cluster is to be found 3.7° SSW from Atria, Alpha Trianguli Australis (mag. 1.9).

Object Description

C107 is a little studied southern globular cluster which, if it was not for the advent of CCD detectors, would perhaps have remained so. These detectors have enabled quite modest telescopes (that is by professional standards!) to produce reliable photometry of it. However, Sarajedini & Costa produced the first deep photometry for nearly 2,700 stars in C107 using only a 0.9 metre telescope in 1991. They used their data to derive an age for C107 by two different methods (theoretical isochrones and differential techniques). They found its age to be similar to M92 and in the order of 16 billion years. Their photometry also revealed a substantial blue straggler star (BSS) population of at least 27 (see C54), rather more concentrated towards the middle than is usual. Amongst these, three were eclipsing binaries, two of which were in contact, which lends weight to the hypothesis that BSS are the result of mass transfer or mergers in binary systems.

C108

Image

Buil-Thouvenot

Database

Name and/or Catalogue Designations:
NGC 4372
Dunlop 67

Type of object: Globular Cluster

Catalogue position for epoch J2000.0
Right ascension: 12h 25m 45.4s
Declination: −72° 39′ 33″

Constellation: Musca

Object information
Magnitude: 7.8
Size: 18.6′
Object classification: 12
NGC Description: pretty faint, large, round, stars 12th to 16th magnitude
Note: contains 8 known SX Phe stars.
Distance from the Sun: 4.9 kpc.
Distance from the galactic centre: 6.9 kpc.

Star atlas chart numbers:
Millennium Star Atlas, Charts 1013–1014, Volume II
Sky Atlas 2000.0, Chart 25
Uranometria 2000, Chart 450, Volume 2

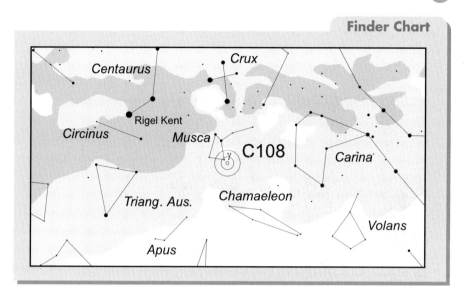

Visual Description

In theory, C108 should be a more visually impressive object than C107 as it is brighter overall and its member stars a touch brighter as well but reality is somewhat different. What counts against it is its larger size and very loose structure. This results in a very low surface brightness and this is of course the most crucial factor affecting visibility. What does help in finding it is a 6.6 magnitude (yellowish) star on or just outside its north-western edge. But conversely this star also hinders visibility and binoculars (even 20 × 80s) struggle with it, only showing the very faintest of smudges to the south-east of the star. An 8 inch (200 mm) telescope does only a little better but the cluster does grow appreciably in size in this aperture, almost reaching as far as the star. C108 does not have a marked central concentration and its member stars are still difficult at this aperture as there seems to be so few of them. Overall not one of best globulars and, if viewing low to the horizon, very difficult. It lies 0.7° south west of Gamma Muscae (mag. 3.84).

Object Description

C108 is a globular cluster with very low metallicity, very low central condensation and is somewhat obscured (reddened). Its one claim to fame is the presence of several ultra-short period Cepheid stars, known as SX Phoenicis variables (SX Phe for short). The detection of these important pulsating variable stars in nearby NGC 5053 (Nemec 1989) prompted the search for more. Kaluzny & Krzeminski (1993) examined C108 looking for short-period variables. They were not disappointed, finding no less than 8 SX Phe stars (plus 8 contact binaries) and also determining their periods. S Phe stars, and other pulsating variables, are important as the existence of a relationship between period and absolute luminosity means that they can provide an independent means of determining distances to the systems that contain them.

C109

Image

Steve Lee

Database

Name and/or Catalogue Designations:
NGC 3195
PK296-20.1

Type of object: Planetary Nebula

Catalogue position for epoch J2000.0
Right ascension: 10h 09m 30.2s
Declination: −80° 51′ 48″

Constellation: Chamaeleon

Object information
Magnitude: 8.4 (Hubble GSC)
Size: 40.0″ × 30.0″
Object classification: 3
NGC Description: remarkable, pretty bright, small, irregular extended, 3 stars nearby
Note: image available in Hubble Planetary Nebula Gallery.
Faint star to west, magnitude 12.5

Star atlas chart numbers:
Millennium Star Atlas, Charts 1023–1024, Volume II
Sky Atlas 2000.0, Chart 25
Uranometria 2000, Chart 465, Volume 2

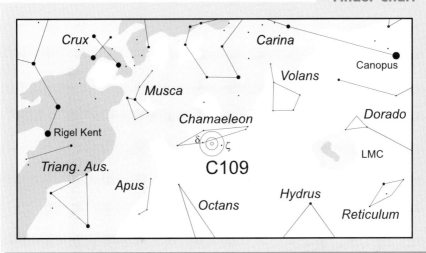

Visual Description

Congratulations if you have got this far but if you are observing the Caldwell objects with binoculars this one is the sting in the tail! C109 is a dim Jupiter-sized planetary nebula discovered by Sir John Herschel in 1835 and his description of "pretty bright" refers to its appearance through his 18 inch telescope. The Hubble Guide Star Catalogue (GSC) catalogued this object as "non-stellar" and magnitude 8.38. However, this seems a gross overestimation as to its visibility and I would suggest magnitude 11 would be a better indication. There is a faint star (magnitude 12.5) on its west. It is therefore probably beyond the reach of binoculars with a 6 inch (150 mm) telescope about the minimum required. As the aperture, and just as importantly the magnification, increases then its nature becomes more obvious and above 12 inches (300 mm) it even has a tinge of blue. Nebula filters (try an O-III) dim the adjacent star and help show its full 40 arc-second extent. The central star however (magnitude 15.5), is all but impossible whatever the aperture. C109 can be found exactly halfway between the 5th magnitude stars Delta1/Delta2 and Zeta Chamaeleontis

Object Description

Although C109 features in the Hubble Planetary Nebula Gallery (credit Howard Bond) unfortunately no associated data has (yet) been published for it. So for the time being we must rely on ground-based observations and in particular two papers. Firstly Meatheringham et al. (1988) studied several southern planetaries, including C109, concentrating mainly on physical properties plus distances. Secondly Kingsburgh & Barlow (1994) looked more closely at elemental abundancies in several southern planetaries. Both used images taken with the Anglo Australian Telescope and findings include: the diameter of the nebula is about half a light year, its distance from us is in the order of 6000 light years and the age of the nebula (as opposed to its parent star) is just over 4000 years. Elements found in the nebula include (in addition to hydrogen and helium of course) oxygen, nitrogen, carbon, neon, argon and sulphur.

Appendix A

The Caldwell Objects and Visibility Limits

Visibility Limit Northern Observer	Caldwell Number	Other Catalogue Number	Common Name or Description	Visibility Limit Southern Observer
	1	NGC 188	very old cluster	
	2	NGC 40		
	3	NGC 4236		
	4	NGC 7023		
	5	IC 342		
	6	NGC 6543	Cat's Eye Nebula	
	7	NGC 2403		
	8	NGC 559		
	9	Sh2-155	Cave Nebula	Rio
	10	NGC 663		
	11	NGC 7635	Bubble Nebula	
	12	NGC 6946		
	13	NGC 457	Owl Cluster or Phi Cass Cluster	Brisbane
	14	NGC 869/884	Double Cluster h and Chi Persei	
	15	NGC 6826	Blinking Nebula	Cape Town
	16	NGC 7243		
	17	NGC 147	M31 Satellite Galaxy	Sydney
	18	NGC 185	M31 Satellite Galaxy	
	19	IC 5146	Cocoon Nebula	Melbourne
	20	NGC 7000	North America Nebula	Wellington

Visibility Limit Northern Observer	Caldwell Number	Other Catalogue Number	Common Name or Description	Visibility Limit Southern Observer
	21	NGC 4449		
	22	NGC 7662		Christchurch
	23	NGC 891		
	24	NGC 1275	Perseus A	
	25	NGC 2419	most distant globular	
	26	NGC 4244		
	27	NGC 6888	Crescent Nebula	
	28	NGC 752		
	29	NGC 5005		
	30	NGC 7331		
	31	IC 405	Flaming Star	
	32	NGC 4631	Whale Galaxy	Port Stanley
	33	NGC 6992/5	Eastern Veil	
	34	NGC 6960	Western Veil	
	35	NGC 4889	in Coma Cluster	
	36	NGC 4559		
	37	NGC 6885		
	38	NGC 4565	Needle Galaxy	
	39	NGC 2392	Eskimo Nebula	
	40	NGC 3626		
	41	Melotte 25	Hyades	
	42	NGC 7006	distant globular	
	43	NGC 7814		
	44	NGC 7479		
	45	NGC 5248		
	46	NGC 2261	Hubble's Variable Nebula	
	47	NGC 6934		
	48	NGC 2775		
	49	NGC 2237–9	Rosette Nebula	
	50	NGC 2244		
	51	IC 1613		

Visibility Limit Northern Observer	Caldwell Number	Other Catalogue Number	Common Name or Description	Visibility Limit Southern Observer
	52	NGC 4697		
	53	NGC 3115	Spindle Galaxy	
	54	NGC 2506		
	55	NGC 7009	Saturn Nebula	
	56	NGC 246		
	57	NGC 6822	Barnard's Galaxy	
	58	NGC 2360		
	59	NGC 3242	Ghost of Jupiter	
	60	NGC 4038		
	61	NGC 4039		
	62	NGC 247		
	63	NGC 7293	Helix Nebula	
	64	NGC 2362	Tau CMa Cluster	
Helsinki	65	NGC 253	Sculptor Galaxy	
	66	NGC 5694		
UK	67	NGC 1097		
Paris	68	NGC 6729	R CrA Nebula	
	69	NGC 6302	Bug Nebula	
	70	NGC 300		
	71	NGC 2477		
	72	NGC 55	galaxy in the Sculptor Group	
	73	NGC 1851		
	74	NGC 3132		
Montreal/Ottawa	75	NGC 6124		
	76	NGC 6231		
Rome	77	NGC 5128	Centaurus A	
New York/Madrid	78	NGC 6541		
Lisbon, Portugal	79	NGC 3201		
Athens	80	NGC 5139	Omega Centauri	
San Francisco	81	NGC 6352		
Tokyo	82	NGC 6193		

Visibility Limit Northern Observer	Caldwell Number	Other Catalogue Number	Common Name or Description	Visibility Limit Southern Observer
Island of Crete	83	NGC 4945		
Atlanta	84	NGC 5286		
	85	IC 2391	Omicron Vel Cluster	
	86	NGC 6397		
	87	NGC 1261		
	88	NGC 5823		
	89	NGC 6087	S Normae Cluster	
	90	NGC 2867		
	91	NGC 3532		
	92	NGC 3372	Eta Car Nebula	
	93	NGC 6752		
	94	NGC 4755	Jewel Box Cluster	
	95	NGC 6025		
Miami	96	NGC 2516		
	97	NGC 3766		
	98	NGC 4609		
	99		Coalsack Dark Nebula	
	100	(IC 2944)	Lambda Cen Cluster	
Hong Kong	101	NGC 6744		
	102	IC 2602	Theta Car Cluster	
	103	NGC 2070	Tarantula Nebula	
	104	NGC 362		
	105	NGC 4833		
	106	NGC 104	47 Tucanae	
	107	NGC 6101		
Barbados	108	NGC 4372		
	109	NGC 3195		

Appendix B

An Index to the Caldwell Objects

Catalogue Number		Name/Description	Caldwell Number
NGC	40		2
NGC	55	galaxy in the Sculptor Group	72
NGC	104	47 Tucanae	106
NGC	147	M31 Satellite Galaxy	17
NGC	185	M31 Satellite Galaxy	18
NGC	188	very old cluster	1
NGC	246		56
NGC	247		62
NGC	253	Sculptor Galaxy	65
NGC	300		70
NGC	362		104
NGC	457	Owl Cluster or Phi Cass Cluster	13
NGC	559		8
NGC	663		10
NGC	752		28
NGC	869/884	Double Cluster h and Chi Persei	14
NGC	891		23
NGC	1097		67
NGC	1261		87
NGC	1275	Perseus A	24
NGC	1851		73
NGC	2070	Tarantula Nebula	103
NGC	2237–9	Rosette Nebula	49
NGC	2244		50
NGC	2261	Hubble's Variable Nebula	46
NGC	2360		58
NGC	2362	Tau CMa Cluster	64
NGC	2392	Eskimo Nebula	39
NGC	2403		7
NGC	2419	most distant globular	25
NGC	2477		71
NGC	2506		54
NGC	2516		96
NGC	2775		48
NGC	2867		90
NGC	3115	Spindle Galaxy	53
NGC	3132		74
NGC	3195		109

Catalogue Number		Name/Description	Caldwell Number
NGC	3201		79
NGC	3242	Ghost of Jupiter	59
NGC	3372	Eta Car Nebula	92
NGC	3532		91
NGC	3626		40
NGC	3766		97
NGC	4038		60
NGC	4039		61
NGC	4236		3
NGC	4244		26
NGC	4372		108
NGC	4449		21
NGC	4559		36
NGC	4565	Needle Galaxy	38
NGC	4609		98
NGC	4631	Whale Galaxy	32
NGC	4697		52
NGC	4755	Jewel Box Cluster.	94
NGC	4833		105
NGC	4889	in Coma Cluster	35
NGC	4945		83
NGC	5005		29
NGC	5128	Centaurus A	77
NGC	5139	Omega Centauri	80
NGC	5248		45
NGC	5286		84
NGC	5694		66
NGC	5823		88
NGC	6025		95
NGC	6087	S Normae Cluster	89
NGC	6101		107
NGC	6124		75
NGC	6193		82
NGC	6231		76
NGC	6302	Bug Nebula	69
NGC	6352		81
NGC	6397		86
NGC	6541		78
NGC	6543	Cat's Eye Nebula	6
NGC	6729	R CrA Nebula	68
NGC	6744		101
NGC	6752		93
NGC	6822	Barnard's Galaxy	57
NGC	6826	Blinking Nebula	15
NGC	6885		37
NGC	6888	Crescent Nebula	27
NGC	6934		47
NGC	6946		12
NGC	6960	Western Veil	34
NGC	6992/5	Eastern Veil	33
NGC	7000	North America Nebula	20
NGC	7006	distant globular	42
NGC	7009	Saturn Nebula	55
NGC	7023		4
NGC	7243		16
NGC	7293	Helix Nebula	63
NGC	7331		30

Catalogue Number		Name/Description	Caldwell Number
NGC	7479		44
NGC	7635	Bubble Nebula	11
NGC	7662		22
NGC	7814		43
IC	342		5
IC	405	Flaming Star	31
IC	1613		51
IC	2391	Omicron Vel Cluster	85
IC	2602	Theta Car Cluster	102
IC	2944	Lambda Cen Cluster	100
IC	5146	Cocoon Nebula	19
Melotte	25	Hyades	41
Sh	Sh2-155	Cave Nebula	9
DN		Coalsack Dark Nebula	99

Appendix C

Observing Certificates

Observing the Caldwell Objects

Silver Certificate

This is to Certify that

....

has observed all

the Caldwell Objects visible from

....

Date: ....

David Ratledge

Observing the Caldwell Objects

Gold Certificate

This is to Certify that

_____

has observed all

109 Caldwell Objects

Date: _____

David Ratledge